microterrors

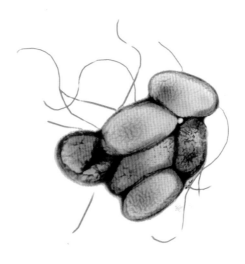

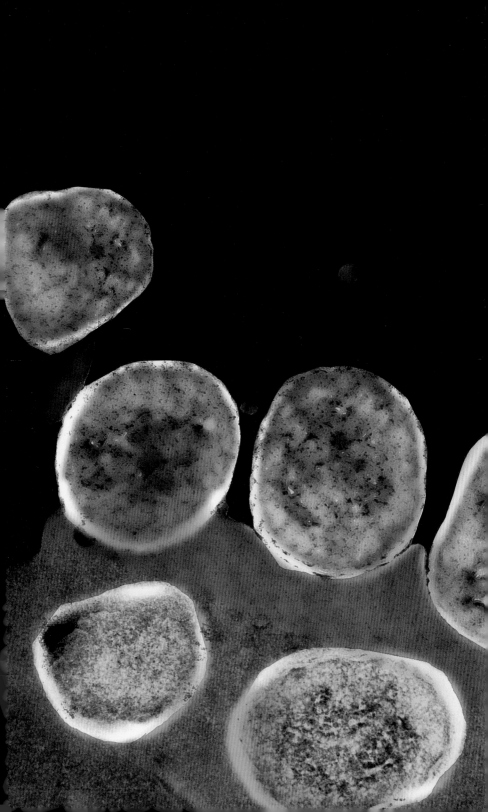

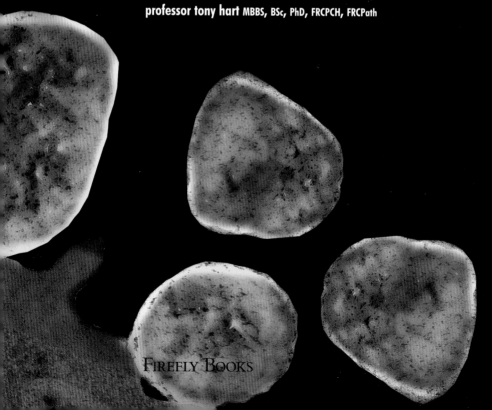

microterrors
the complete guide to bacterial, viral and fungal
infections that threaten our health

professor tony hart MBBS, BSc, PhD, FRCPCH, FRCPath

FIREFLY BOOKS

A FIREFLY BOOK

Published by Firefly Books Ltd. 2004

First printing

Publisher Cataloging-in-Publication Data (U.S.)

Hart, C. A. (Charles Anthony).

　　Microterrors : the complete guide to bacterial, viral and fungal infections that threaten our health /

　　Tony Hart.

[192] p. : col. photos. ;　cm.

Includes index.

Summary: Profiles of both naturally occurring and bioengineered bacteria, viruses, fungi, and protozoa. Includes date of discovery, place of origin, period of incubation, symptoms, treatments, and other key facts.

ISBN 1-55297-971-7

ISBN 1-55297-970-9 (pbk.)

1. Microorganisms. 2. Microbiology -- Popular works.
I. Title.

579　dc22　QR56.H37　2004

National Library of Canada Cataloguing in Publication

Hart, C. A. (Charles Anthony)

Microterrors : the complete guide to bacterial, viral and fungal infections that threaten our health /

　　Tony Hart.Includes index.

ISBN 1-55297-971-7 (bound).--ISBN 1-55297-970-9 (pbk.)

1. Medical microbiology--Handbooks, manuals, etc. I. Title.

QR56.H37 2004　　616.9'041　　C2004-902334-9

Published in the United States in 2004 by
Firefly Books (U.S.) Inc.
P.O. Box 1338, Ellicott Station
Buffalo, New York 14205

Published in Canada in 2004 by
Firefly Books Ltd.
66 Leek Crescent
Richmond Hill, Ontario L4B 1H1

Conceived and created by
Axis Publishing Limited
8c Accommodation Road
London NW11 8ED

Creative Director: Siân Keogh
Editorial Director: Anne Yelland
Design: Axis Design Editions
Managing Editor: Conor Kilgallon
Production Manager: Toby Reynolds

Printed in Thailand

contents

INTRODUCTION: what are microbes?

They are invisible to us, but each day we are exposed to a multitude of different microorganisms. For the most part they do us no harm, but occasionally we are challenged by disease-causing (pathogenic) microbes that can be detrimental to our health.

All the microorganisms that live on us and in our environment are at the lowest segment of the trunk of the evolutionary tree. Yet what they lack in sophistication and superior development, they overcome by tremendous versatility, ability to multiply rapidly, and, for some microbes, an incredible potential to mutate.

A WEALTH OF KNOWLEDGE

Prior to 1675, nothing was known of microorganisms. It was the invention of the microscope by the Dutchman Antonie van Leeuwenhoek that allowed us to see these previously invisible "animalcules" as he called them. With the light microscope he was able to see bacteria and larger microorganisms. Although there was evidence that smaller microorganisms existed and caused disease in humans and other animals, it was not until the development of the electron microscope in 1939 that viruses could be seen. Since then, our understanding of the classification diversity and pathogenic potential of the viruses, bacteria, fungi, and protozoa that infect and coexist with humans has increased exponentially. Alongside this knowledge, it is important to stress that new pathogens continue to emerge and old pathogens reemerge or evolve to become even more pathogenic.

VIRUSES

Viruses are the most primitive of the microbes. They can have no existence independent of their host. To reproduce themselves, they must first attach to the host cell. This they do by recognizing the particular shape of a structure on the cell, which acts as a receptor. If this receptor is not present, the cell cannot be infected.

Once the virus has attached to the host cell it enters it and "uncoats." By various mechanisms, it then shuts off the host cell's normal biosynthetic pathways and subverts them into making new copies of the virus. These assemble either as large clusters inside the cell or at the cell membrane and are released by the cell bursting open, or by the virus "budding" from the cell surface. Each infected cell can release thousands of new virus particles, which then go on to infect other cells. This makes viruses "obligate intracellular parasites."

Finally, the host cell dies. If this process affects lots of cells, tissue damage occurs, which can be permanent or fatal.

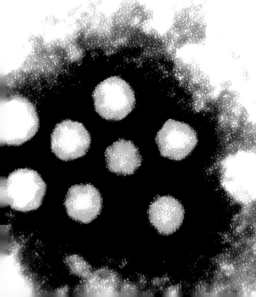

The astrovirus, page 26.

However, the infected cells are not passive, but cry for help by releasing small molecules called cytokines and chemokines to recruit our immune system to the site of the infection. This contributes to the symptoms of infection, such as fever, and, in the respiratory tract for example, partial blockage of the airways and the production of sputum.

There are several consequences resulting from the virus only being able to reproduce itself inside a host cell. Firstly, once it has been excreted from the host, virus numbers fall (this is not the case for most bacteria which are free-living). Secondly, because the virus uses our synthetic machinery to reproduce itself, designing effective drugs to inhibit viruses is very difficult and this is why we have so few antiviral drugs.

VIRUS CHARACTERISTICS

Viruses are very small, ranging in size from 25 to 350 nanometers (nm) in diameter. Some just have a protein coat, a "capsid," but others have an envelope of "lipid" surrounding the capsid. This envelope is usually acquired as the virus buds from the cell it has infected and is thus derived from the host cell membrane. In general, the viruses that have envelopes survive less well in the outside world (the environment) and are easily destroyed.

Viruses are unusual in that for some, the genetic code is of ribonucleic acid (RNA) and for others it is deoxyribonucleic acid (DNA), but never both, unlike humans. The shape of the capsid for some viruses is helical (spiral) and for others, it is a structure with 20 "facets." These characteristics make it possible to classify viruses into different families, genera, and species to provide a universally accepted scheme for naming and describing new viruses.

The influenza virus, pages 30–31.

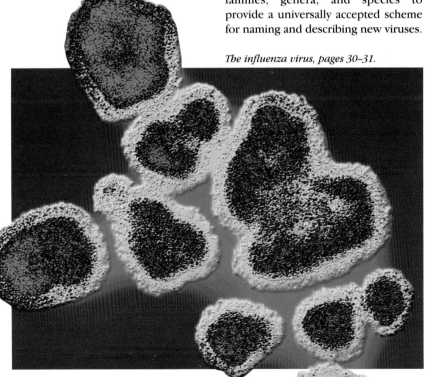

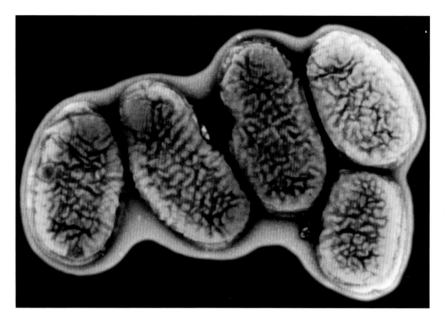

BACTERIA

Although some bacteria are obligate intracellular parasites like viruses, most are free-living and, providing they can scavenge enough food, will keep on reproducing. They reproduce by binary fission, each bacterium dividing to produce two identical offspring (progeny). Under optimal conditions, they can divide every 20 minutes so that after 18–24 hours, one bacterial cell can divide to produce up to 10 million descendants. To put this in context, the world's human population is around 5,000 million.

BACTERIA CHARACTERISTICS

Bacteria range in size from 0.5 to 1 micron (μ) in diameter to 3–15 μ long. Their classification is based upon their shape, staining characteristics, growth characteristics, biochemical properties and, increasingly, on their genomic make up. Bacteria can be rod shaped (bacilli), spherical (cocci), or spiral (spirochaetes). The bacilli can be straight or bent. Cocci can be arranged in pairs (diplococci), in clusters

The melioidosis bacteria, page 148.

(staphylococci) or in long chains (streptococci). Some bacteria, such as *Bacillus anthracis* (which causes anthrax), produce survival packages called spores. These allow the bacteria to survive under harsh conditions such as drying, heating, and irradiation, where non-spore producing bacteria are killed. When conditions improve, the spores germinate and the bacterium reproduces itself.

In 1895, Dr. Christian Gram, a Danish physician, devised a staining technique that is still used today to classify bacteria. Gram-positive bacteria, which stain blue, have one cell membrane and a thick cell wall. Gram-negative bacteria, which stain pink or red, have two cell membranes and a thinner cell wall.

Some bacteria are killed in the presence of oxygen. These are called obligate anaerobes. Other bacteria (facultative anaerobes) can grow whether oxygen is present or not.

FUNGI

The fungi form a large kingdom, but only a small number are pathogenic for humans. Fungi differ from bacteria in several ways. They are "eukaryotes," meaning that they have chromosomes in a nucleus and have internal structures called organelles. They are usually bigger than bacteria, more likely to be branching, and reproduce both sexually and by binary fission. Some, such as *Candida albicans*, remain as single cells, whereas others form large branching networks called mycelia.

Fungi can be subdivided into those that cause superficial infections (such as dermatophytes which cause an infection on the skin), those that cause infections in the tissues beneath the skin (subcutaneous), and those that disseminate around the body (systemic). They range from single-celled yeasts, such as *Candida albicans* (mentioned above) to large multicellular organisms with a sex life, such as those causing ringworm and athlete's foot. They have a cell wall and range in size from 5 μ to a few millimeters (mm) long.

The cryptococcosis fungus, page 175.

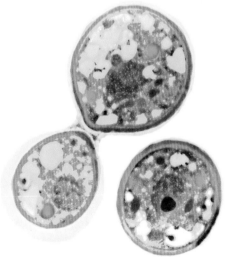

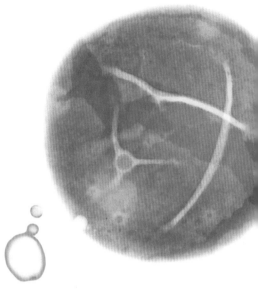

The cryptosporidiosis protozoon, page 180.

PROTOZOA

The protozoa form a large kingdom of eukaryotes consisting of single-celled organisms. They have nuclei and intracellular organelles but no cell wall. These are single-celled organisms and begin to resemble human cells. Some, such as *Entamoeba histolytica*, reproduce by binary fission but others, such as *Cryptosporidium parvum* or *Plasmodium falciparum*, have a sexual phase of reproduction. They are quite fragile and those that are shed into the external environment usually produce hardy, thick-walled cysts. Only a minority of the protozoa are pathogenic for humans. They range in size from 5 to 20 μ.

MULTICELLULAR PARASITES

Nematodes (round worms), trematodes (flukes), and cestodes (tapeworms) consist of large numbers of cells organized together. Some are pathogenic for humans.

WHAT IS NORMAL?

It is estimated that a human adult is less than 10 percent human. An adult comprises some 100 million million (10^{14}) cells but fewer than 10 million million (10^{13}) are human. The remaining 90 percent are the viruses, bacteria, fungi, protozoa, worms, and insects that make up our normal flora.

However, normal flora are not uniformly distributed throughout the body. Microorganisms are found on exterior body surfaces, such as the skin and conjunctivae, and internal body surfaces that communicate with the exterior, such as the mouth, gastrointestinal tract, urethra, and vagina. The blood, organs such as the brain and liver, and respiratory tract below the vocal cords are normally sterile and finding bacteria in these areas usually indicates infection.

BACTERIA, VIRUSES, FUNGI, WORMS

Bacteria make up the largest part of our normal flora. For example, half the wet weight of feces is made up of bacteria. It is estimated that there are approximately 1 million million (10^{12}) bacteria per gram of feces. Some of the bacteria we carry can cause disease if they move from their normal colonization sites. For example, up to 50 percent of us carry *Neisseria meningitidis* in our throats. If it moves to the bloodstream or brain, it causes meningococcal septicemia or meningitis respectively.

Viruses can be found in the absence of infection; for example, once we have been infected by any one of the eight human herpes viruses, that virus will stay in our cells for life. Retroviruses make up three percent of our chromosomes and play an essential role in rearranging and activating our genes.

All of us carry fungi on our skins and yeasts such as *Candida albicans* can be found in the mouths and intestines of a large proportion of humans. We also carry protozoa such as *Trichomonas oralis* and *Entamoeba coli* in the mouth and colon respectively.

We can even be hosts for worms such as the threadworm (*Enterobius vermicularis*) and whipworm (*Trichuris trichiura*), without knowing it. We all also act as hosts for insects such as the follicle mite (*Demodex follicularis*), which lives in hair follicles, and the louse (*Pediculus humanus*).

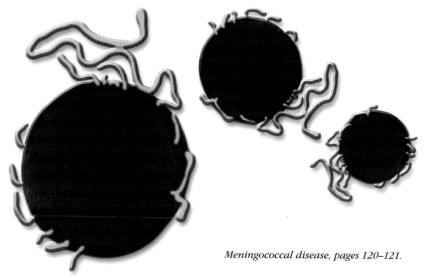

Meningococcal disease, pages 120–121.

ROLE OF NORMAL FLORA

Our normal flora is for the most part for our benefit, but occasionally it can cause problems. For example, if a patient has an operation on the colon, the colonic bacteria, in particular anaerobic bacteria such as *Bacteroides fragilis*, spill into the peritoneum and cause peritonitis and abscesses. This risk can be minimized by giving antibiotics to cover the operating period.

The lack of normal flora or the wrong flora, however, poses greater problems. In experiments, if rodents are maintained without a normal flora, their immune system does not develop well and they are less able to mount a defense against invading pathogens. In addition, normal intestinal flora is important in the proper development of the gut. In germ-free rats, the intestine is greatly decreased in weight and surface area and the liver is decreased in size. Normal flora is also important in synthesizing some vitamins, for example vitamins K and the B complex.

Finally, it is well recognized that perturbations of the intestinal flora, for example during treatment with certain antibiotics, will cause diarrhea. So overall, the normal flora is beneficial and its absence is detrimental.

BODY SITES NORMALLY COLONIZED

GASTROINTESTINAL TRACT	SKIN
The **gastrointestinal tract** is the major site for colonization and there are an estimated 2–4lb (1–2kg) of bacteria in this part of the adult body. There is also a great diversity of bacterial species here, particularly in the colon. It is estimated that 500–1,000 different species live in the gut. Indeed, it is known that the metabolic activity of the bacteria in our intestines greatly exceeds our own metabolism and the genetic diversity of these bacteria is immense.	The **skin** surface area of an average adult human is some 18ft^2 (1.7m^2). Approximately every 3–4 days we lose a layer of skin. However, our skin is not shed in one piece like a snake's skin, but by loss of individual cells or small groups of cells. These are called squames and attached to the squames are bacteria. Shed squames could, for example, pose a risk of infection if they were to fall into an open wound during a surgical operation. To avoid this, surgeons wear tightly woven surgical gowns that prevent their own shed squames from passing through and into the operating site.

Division and mutation

Under ideal conditions, bacteria divide every 20 minutes and it is estimated that the mutation rate is, at lowest, one in every 100 million (10^8) cells. If there are approximately 100 million million (10^{14}) bacteria present, it follows that in every cell division (every 20 minutes), one million mutant bacteria will arise. It is clear from this that the normal flora and, in particular, the intestinal bacteria, have a great capacity to react to any adverse conditions to which they are exposed. This includes the administration of antibiotics to fight infection and it is no coincidence that the intestinal bacteria are a major reservoir of antibiotic resistance genes.

Distribution

The skin microorganisms are not uniformly distributed over the surface. Moist areas, such as the forehead or back, have larger numbers of bacteria (in the order of tens of thousands per square inch). By contrast, our hands are scantily colonized, by perhaps 600–1,200 per square inch (100–200 per square centimeter).

The normal skin flora serve a useful role in killing any exogenous bacteria that are picked up by touching. However the hands, which are relatively deficient in normal flora, are less able to clear exogenous bacteria and they are thus one of the most important modes of transfer of bacteria. This is why such emphasis is placed on handwashing by healthcare staff.

how to use this book

The book is organized around the four main groups of potentially pathogenic organisms: viruses, bacteria, fungi, and protozoa. Within these broad groupings, the organisms are broken down into families, and then individual species. In general, one organism is featured on a page, but where an organism is particularly important, two pages are devoted to its description.

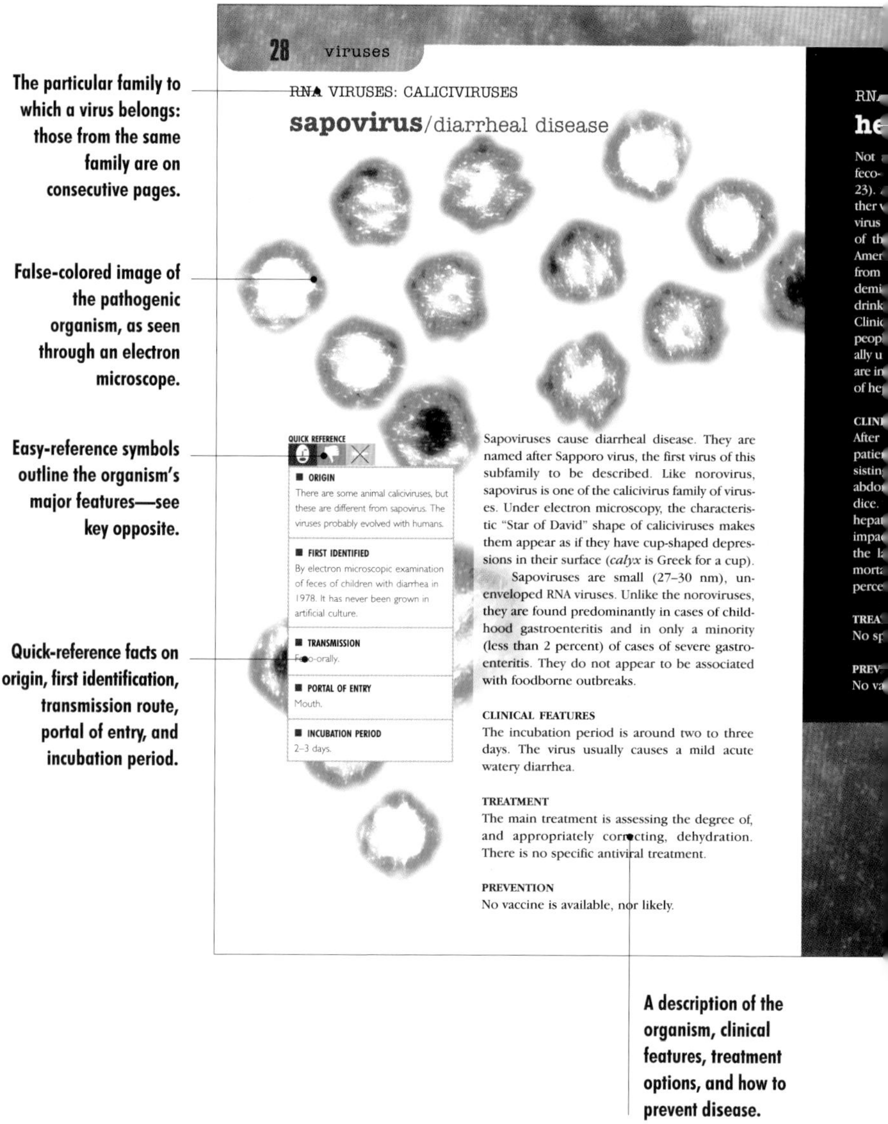
The particular family to which a virus belongs: those from the same family are on consecutive pages.
False-colored image of the pathogenic organism, as seen through an electron microscope.
Easy-reference symbols outline the organism's major features—see key opposite.
Quick-reference facts on origin, first identification, transmission route, portal of entry, and incubation period.
A description of the organism, clinical features, treatment options, and how to prevent disease.

28 viruses

RNA VIRUSES: CALICIVIRUSES

sapovirus/diarrheal disease

QUICK REFERENCE

■ ORIGIN
There are some animal caliciviruses, but these are different from sapovirus. The viruses probably evolved with humans.

■ FIRST IDENTIFIED
By electron microscopic examination of feces of children with diarrhea in 1978. It has never been grown in artificial culture.

■ TRANSMISSION
Feco-orally.

■ PORTAL OF ENTRY
Mouth.

■ INCUBATION PERIOD
2–3 days.

Sapoviruses cause diarrheal disease. They are named after Sapporo virus, the first virus of this subfamily to be described. Like norovirus, sapovirus is one of the calicivirus family of viruses. Under electron microscopy, the characteristic "Star of David" shape of caliciviruses makes them appear as if they have cup-shaped depressions in their surface (*calyx* is Greek for a cup).

Sapoviruses are small (27–30 nm), unenveloped RNA viruses. Unlike the noroviruses, they are found predominantly in cases of childhood gastroenteritis and in only a minority (less than 2 percent) of cases of severe gastroenteritis. They do not appear to be associated with foodborne outbreaks.

CLINICAL FEATURES
The incubation period is around two to three days. The virus usually causes a mild acute watery diarrhea.

TREATMENT
The main treatment is assessing the degree of, and appropriately correcting, dehydration. There is no specific antiviral treatment.

PREVENTION
No vaccine is available, nor likely.

The organism's name, followed by the condition or disease which it causes.

The major biological grouping to which an organism belongs: viruses, bacteria, fungi, or protozoa.

THE ICONS

For ease of reference, a series of icons has been devised to provide an at-a-glance key to an organism's major features. These are illustrated and described below. These always appear in the same order, namely whether the disease caused can be fatal; the most common transmission routes, such as by hand, in food and drink, or by insects or animals; and the treatment and prevention options, including whether an effective vaccine is available or not.

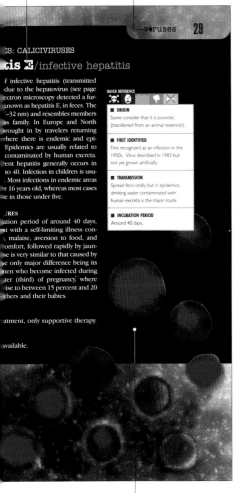

viruses 29

S: CALICIVIRUSES

tis E/infective hepatitis

f infective hepatitis (transmitted due to the hepatovirus (see page ectron microscopy detected a fur-known as hepatitis E, in feces. The -32 nm) and resembles members s family. In Europe and North rought in by travelers returning here there is endemic and epi-Epidemics are usually related to contaminated by human excreta. ent hepatitis generally occurs in to 40. Infection in children is usu-. Most infections in endemic areas r 16 years old, whereas most cases e in those under five.

QUICK REFERENCE

■ **ORIGIN**
Some consider that it is zoonotic (transferred from an animal reservoir).

■ **FIRST IDENTIFIED**
First recognized as an infection in the 1950s. Virus described in 1983 but not yet grown artificially.

■ **TRANSMISSION**
Spread feco-orally but in epidemics, drinking water contaminated with human excreta is the major route.

■ **INCUBATION PERIOD**
Around 40 days.

JRES
ation period of around 40 days, t with a self-limiting illness con-, malaise, aversion to food, and omfort, followed rapidly by jaun-se is very similar to that caused by nen who become infected during ter (third) of pregnancy, where ise to between 15 percent and 20 thers and their babies.

atment, only supportive therapy.

vailable.

Organisms on a black background prove fatal in a significant number of cases.

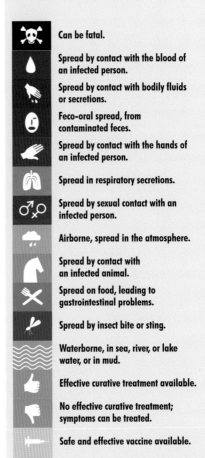

☠	**Can be fatal.**
💧	**Spread by contact with the blood of an infected person.**
🖐	**Spread by contact with bodily fluids or secretions.**
👤	**Feco-oral spread, from contaminated feces.**
✋	**Spread by contact with the hands of an infected person.**
🫁	**Spread in respiratory secretions.**
♂♀	**Spread by sexual contact with an infected person.**
🌧	**Airborne, spread in the atmosphere.**
🐴	**Spread by contact with an infected animal.**
✗	**Spread on food, leading to gastrointestinal problems.**
🦟	**Spread by insect bite or sting.**
〰	**Waterborne, in sea, river, or lake water, or in mud.**
👍	**Effective curative treatment available.**
👎	**No effective curative treatment; symptoms can be treated.**
💉	**Safe and effective vaccine available.**
✕	**No effective vaccine available.**

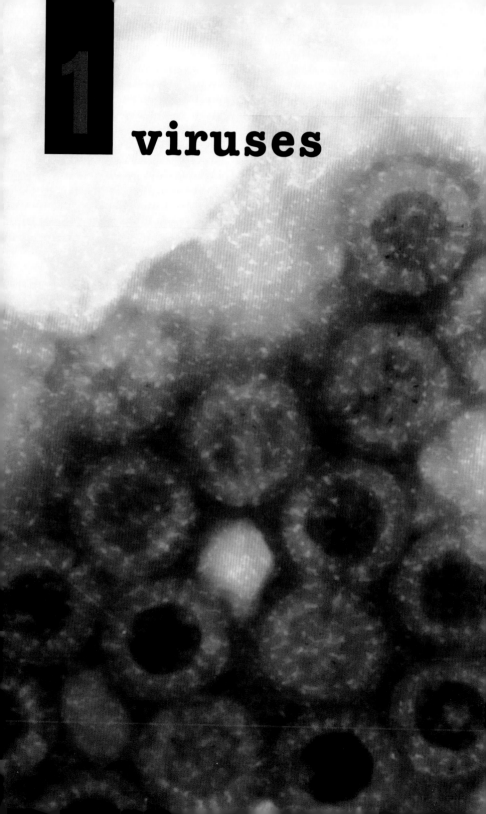

1 viruses

Viruses are usually harmless or mild. However, some can be devastating. They are the most primitive of microbes and are incapable of existing without a host. Once inside a cell, thousands of virus particles then spread to other cells, killing the host cells in the process and subsequently damaging the body.

Viruses are easily destroyed by disinfectants when outside the body but are difficult to eliminate, even with antiviral drugs, once infection has taken place.

PRION DISEASES
kuru/
transmissible spongiform encephalopathy

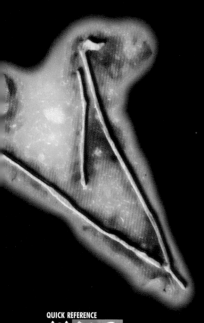

Kuru was first described in 1957 when an outbreak of a disease characterized by unsteady gait, loss of balance, and progression to death was found among a tribal group (the Fore Language Group) in Papua New Guinea. Kuru means "shivering" in the Fore language, and the disease particularly affected children and females.

Microscopic examination of the brains of the deceased showed a sponge-like appearance with lots of holes and very little inflammation. The disease could be transmitted to laboratory animals by inoculating them with brain tissue from patients who had died from Kuru.

INFECTION THROUGH CANNIBALISM
It is thought that the disease was spread by ritual cannibalism. When a member of the tribe died, other members believed that the spirit of the deceased could be maintained by cooking and eating the body. Women and children tended to be left with offal, such as the brain, which explains the higher prevalence of disease in these groups. The brain contains the highest concentrations of the infectious agent, which is not itself a virus but a self-replicating protein called a prion.

CLINICAL FEATURES
The shortest incubation period is four years; the longest more than 30 years. The initial symptoms consist of unsteady gait and loss of balance, followed by increasing dementia (loss of short-term memory), coma, and progression to death within a year.

TREATMENT
No treatment is available.

PREVENTION
A ban on ritual cannibalism came into force in 1960 and Kuru has not occurred in any tribe member born after the ban.

QUICK REFERENCE

■ ORIGIN
Probably originally a normal protein in which a mutation had occurred, which was then transmitted by eating affected brains. Between 1957 and 1982, there were a total of 2,584 cases of Kuru.

■ FIRST IDENTIFIED
First case in 1957. Shown to be transmissible in 1960. The prion hypothesis was accepted when Stanley Prusiner received the Nobel Prize for his work.

■ TRANSMISSION
Cannibalism.

■ PORTALS OF ENTRY
Mouth and gastrointestinal tract.

■ INCUBATION PERIOD
4–35 years.

PRION DISEASES

creutzfeldt-jakob disease/
transmissible spongiform encephalopathy

This disease was first described by two German neuropathologists, Creutzfeldt and Jakob. Originally, it was thought to be hereditary and some cases still occur within families. However, accidental transmission to humans has occurred through neurosurgery techniques and also via injections of hormones prepared from the pituitary glands of deceased donors. The disease is therefore transmissible from person to person and experimentally to animals.

It leads to cerebral degeneration and on microscopic examination the brain shows a large number of holes, or "vacuoles," in which the prion protein has accumulated. The incidence of CJD is one case per million population per year worldwide.

CLINICAL FEATURES
Dementia, which evolves to loss of balance, coma, and, often just a year later, death.

TREATMENT
No treatment is available.

PREVENTION
Screening donated tissue and hormones.

QUICK REFERENCE

■ **ORIGIN**
Probably normal brain proteins that have mutated to become nondegradable prion proteins which accumulate in the brain, causing gradual loss of neurons.

■ **FIRST IDENTIFIED**
The disease was first identified in 1927, and was transmitted to animals by direct inoculation of an affected human brain in the 1970s.

■ **TRANSMISSION**
Has been transmitted accidentally by infected brain material either by direct inoculation into the brain or by intramuscular injection.

■ **PORTAL OF ENTRY**
The brain.

■ **INCUBATION PERIOD**
1–40 years.

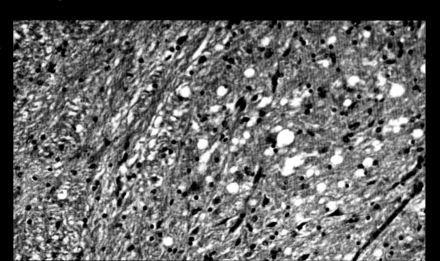

PRION DISEASES
BSE/bovine spongiform encephalopathy

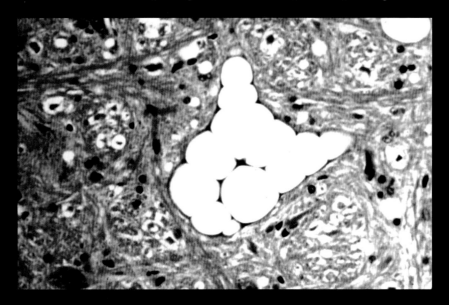

■ **ORIGIN**
Unknown, but hypotheses are that it spread from downers, scrapie (a sheep disease originating in 1738), or mutated cow-brain protein.

■ **FIRST IDENTIFIED**
1985.

■ **TRANSMISSION**
Through contaminated protein pellets.

■ **PORTAL OF ENTRY**
Gastrointestinal tract.

■ **INCUBATION PERIOD**
Shortest around one year, longest not yet known.

The first U.K. case of BSE, or "mad cow disease," was described in 1985. It achieved epidemic proportion in 1992 with over 35,000 cases, and is thought to have arisen by feeding protein pellets made from the carcasses of sheep, cattle, and pigs to cattle. A similar disease (called downers) had also been recorded in the U.S.A.

It is probable that among the carcasses used to make the pellets were cows that had died from a transmissible encephalopthy. Therefore, as more cows died from BSE, more infected carcasses were used to make pellets, amplifying the disease. BSE also spread to domestic cats and some zoo animals, raising the fear that it might spread to humans (see vCJD opposite).

CLINICAL FEATURES
Cattle appear terrified, hypersalivate, and stagger.

TREATMENT
No treatment is available.

PREVENTION
Ban on food pellets made from animal remains.

PRION DISEASES
variant CJD/human "mad cow disease"

Soon after BSE emerged, fears were raised that it might cross the species barrier and infect humans. In this context, it is important to realize that cattle brain was used as extra protein by food-processing companies and some beefburgers contained up to 0.3 oz (1 g) of brain material. Such concerns eventually led to a ban on the use of cattle offal for human consumption. Originally, this ban covered the brain and spinal cord but was later extended to lymphoid tissue such as tonsils, lymph nodes, and thymus.

NEW CASES
Fears grew when a number of cases of new variant CJD (now called variant or vCJD) were described in the U.K. medical journal *The Lancet*. These cases differed from those of classical CJD in that they occurred in younger patients (under the age of 40) and progressed to death at a slower rate. Examination of the brains showed large collections of prion protein in a pattern reminiscent of Kuru (see page 16), which is also transmitted by ingestion.

Analysis of the vCJD prion protein showed that it was the same as the BSE prion protein in key regions (though there were differences in other areas). It is now accepted that the BSE prion protein has caused vCJD.

CLINICAL FEATURES
Very similar to classical CJD except that it occurs in young as well as elderly patients.

TREATMENT
Experimental treatment with antimony compounds, which slows progression.

PREVENTION
Ensuring the safe production of meat products.

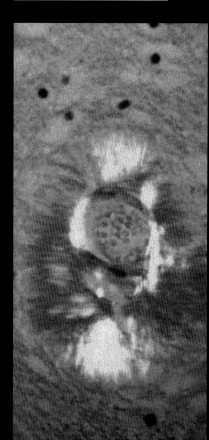

RNA VIRUSES: PICORNA
rhinoviruses

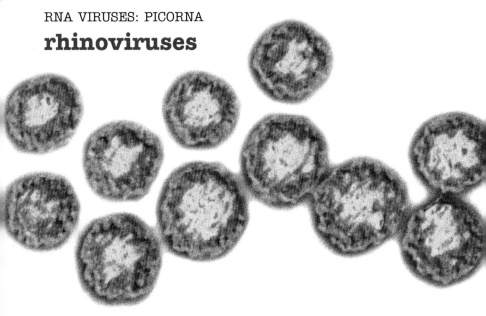

■ **ORIGIN**
Probably evolved with humans.

■ **FIRST IDENTIFIED**
First demonstration of virus (filterable
agar) in 1919; virus first grown in 1959.

■ **TRANSMISSION**
Inhalation of respiratory secretion
but transfer on hands is probably
more significant.

■ **PORTALS OF ENTRY**
Via the upper respiratory tract
and conjunctivae.

■ **INCUBATION PERIOD**
2–3 days.

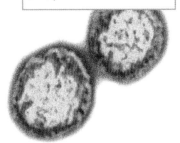

This family of viruses is one of the causes of the
"common cold." Its name is derived from the
Greek for nose (*rhinos*). There are over 100 different rhinoviruses, enough to give each of us a
cold for each year of our lives. It has been estimated that colds are associated with 161 million person days of restricted activity, 26 million
days off school, 25 million days off work, and
27 million visits to doctors each year in the
U.S.A. alone.

CLINICAL FEATURES
The disease presents with a runny nose (known
as *coryza*), cough, sneezing, sore throat, and a
fever. It is usually short-lived (three to five days)
but it can precipitate asthmatic attacks and exac-
erbate chronic bronchitis in elderly patients.

TREATMENT
No reliable antiviral therapy. However, there are
more than 800 common-cold remedies available that provide relief from the symptoms of
the disease. Although these make the patient
feel better, many increase virus growth and
thus aid the spread of the virus to others.

PREVENTION
No safe and effective vaccine is available.

RNA VIRUSES: PICORNA

enteroviruses

The enteroviruses were named after their portal of entry, the intestinal tract (*enteron* is the Greek for intestine). There are more than 95 different enteroviruses subdivided into a vast array of subgroups including polioviruses (1, 2, and 3) and hepatovirus (see pages 22–23), coxsackieviruses (named after the town in the U.S.A. where the first cases were described), *echo*viruses (*e*nteric *c*ytopathic *h*uman *o*rphan), and parechoviruses.

All demonstrate the so-called "iceberg phenomenon." For every 100 patients infected with enteroviruses, only between one and five will develop the disease. The virus can therefore infect a large proportion of the population and circulate undetected. This makes understanding the disease process and its epidemiology difficult, and prevention extremely hard.

CLINICAL FEATURES

Most infections are asymptomatic, and different enteroviruses cause different disease manifestations. These include sore throat, skin rashes, conjunctivitis, myocarditis (infection of the heart muscle), meningitis, and encephalitis (infection of the brain). Hand, foot, and mouth disease (HFMD) presents with blisters on the palms, soles of the feet, and palate.

Since 1997, there have been large outbreaks of HFMD in Malaysia, Taiwan, and Australia due to enterovirus 71, in which there was also encephalitis that paralyzed breathing, causing heart failure and death. Other viruses of the group, such as coxsackie B, can cause Bornholm disease in which there is inflammation of the muscles between the ribs and the diaphragm that is so painful it is called Devil's Grip.

TREATMENT

Most infections do not require treatment. However, for severe infection, such as encephalitis, an antiviral drug (Pleconaril) is available.

PREVENTION

Vaccines are available only for poliovirus and hepatovirus.

QUICK REFERENCE

■ **ORIGIN**
Probably evolved with humans.

■ **FIRST IDENTIFIED**
Late 1940s and early 1950s.

■ **TRANSMISSION**
Feco-orally, either directly or indirectly in food or water.

■ **PORTALS OF ENTRY**
Mouth, pharynx, and intestine.

■ **INCUBATION PERIOD**
3–21 days depending on the particular virus.

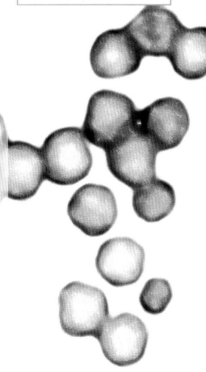

RNA VIRUSES: PICORNA

poliovirus/polio

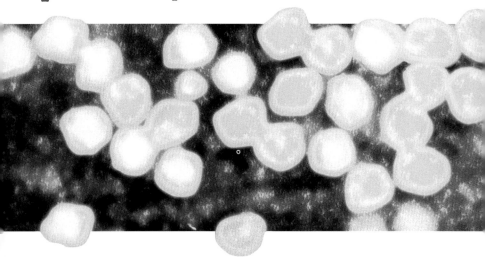

QUICK REFERENCE

■ **ORIGIN**
A description of a priest with symptoms characteristic of polio appears on an Egyptian tomb from 1300 B.C.

■ **FIRST IDENTIFIED**
Filterable agent first demonstrated in 1909, and the virus grown in 1949.

■ **TRANSMISSION**
Feco-orally or oral-to-oral, spread directly or indirectly in food and water.

■ **PORTALS OF ENTRY**
Mouth and conjunctivae.

■ **INCUBATION PERIOD**
3–20 days.

Each of the three different polioviruses (1, 2, and 3) can cause poliomyelitis (polio). The virus grows in the tonsils and mucosa of the upper gastrointestinal tract. During the initial phase, it enters the bloodstream and in some cases, travels to where nerves join muscles. It then ascends the nerve to the spinal cord, causing paralysis. It is not clear why only some patients develop this "flaccid paralysis." However, youth, malnutrition, physical exertion, and pregnancy increase the risk of paralytic polio.

CLINICAL FEATURES
Infection is unapparent in between 90 percent and 95 percent of patients; between 4 percent and 8 percent show signs of upper respiratory tract infection, and between 1 percent and 2 percent have meningitis. Only up to 2 percent develop paralytic polio.

TREATMENT
No specific antiviral therapy is available.

PREVENTION
Two vaccines—Salk, which is given by injection; and Sabin, given orally. The World Health Organization aims to eradicate the polio virus by 2005.

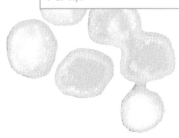

RNA VIRUSES: PICORNA

hepatovirus/infective hepatitis

Hepatovirus was previously called hepatitis A. It causes inflammation of the liver (*hepa* in Greek), and is solely a human pathogen. Infection is common in developing countries where most children will have been infected by the age of 10. In developed countries, where there are good sewage treatment systems, most individuals reach adulthood having never been infected, and therefore epidemics can occur.

As with many infectious diseases, hepatovirus infection is more severe in adults than children. In fact, most children have only asymptomatic infection. Epidemics can occur from eating contaminated food, but the spreading of the infection from person to person is also significant.

CLINICAL FEATURES
Fever, muscle pain, nausea, aversion to food, vomiting, and sometimes pain around the liver. The patient turns yellow (known as jaundice).

TREATMENT
No specific treatment is available.

PREVENTION
The inactivated whole virus vaccine is effective, especially if boosters are given.

QUICK REFERENCE

■ **ORIGIN**
Probably evolved with humans.

■ **FIRST IDENTIFIED**
First recognized as infectious during World War II, and in 1973, virus particles were identified in the feces of infected patients.

■ **TRANSMISSION**
Feco-orally, directly or indirectly in water or food.

■ **PORTAL OF ENTRY**
Mouth.

■ **INCUBATION PERIOD**
2–6 weeks (1 month on average).

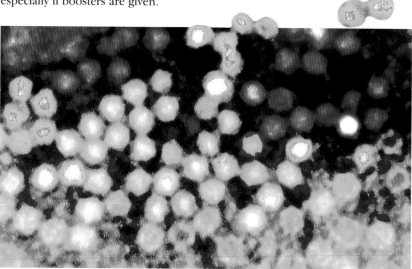

RNA VIRUSES: REOVIRUS

rotavirus/diarrheal disease

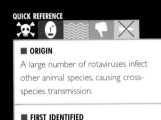
Diarrheal disease is the fourth most common cause of death in developing countries, responsible for over two million deaths each year, predominantly in children under five years of age. More than 45 different viruses, bacteria, and protozoa can cause diarrheal disease, but rotavirus is one of the most significant causes in infants.

All of us will be infected at some time in our lives, often more than once. However, it is only the first two or three episodes of infection that are symptomatic. Between 20 percent and 60 percent of children hospitalized with gastroenteritis are infected with rotavirus and worldwide, rotavirus is estimated to cause more than 500,000 deaths each year.

The virus has a characteristic wheel-like shape (*rota* is Latin for wheel). An infected infant excretes approximately 100,000 million viruses per milliliter of stool. An infective dose contains as few as 100 viruses, making transmission very easy. In temperate countries, rotavirus infections increase in the winter.

MUTATIONS
The rotavirus is highly mutable and can change its surface proteins with ease. Most disease is due to group A rotaviruses, but group B rotaviruses have emerged in China and more recently in India, causing enormous epidemics that have affected adults as well as children.

CLINICAL FEATURES
After a short incubation period (two to three days), infants present with acute voluminous watery diarrhea, and possibly vomiting. Patients can be feverish ("febrile"), though this is not a major characteristic. On average the illness lasts between six and seven days, but relapse can occur with further episodes of diarrhea.

TREATMENT
The cause of death from diarrheal disease is dehydration, and rotavirus is a particularly potent cause of dehydration. It is therefore vital

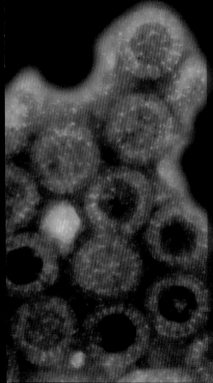

to assess the degree of dehydration and appropriate rehydration, either with oral rehydration solutions (water, salts, and glucose) or the intravenous infusion of fluid. There is no specific antiviral therapy, but administration of probiotic bacteria such as *Lactobacillus casei* can ameliorate the disease.

PREVENTION
A live oral vaccine introduced in the U.S.A., which in trials had worked very well in both developed and developing countries, was later linked to a small number of cases of bowel obstruction (intussusception) and therefore withdrawn. Fortunately, at least four new vaccines are currently undergoing clinical trials.

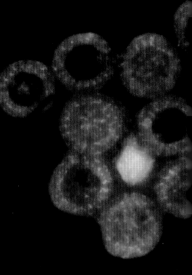

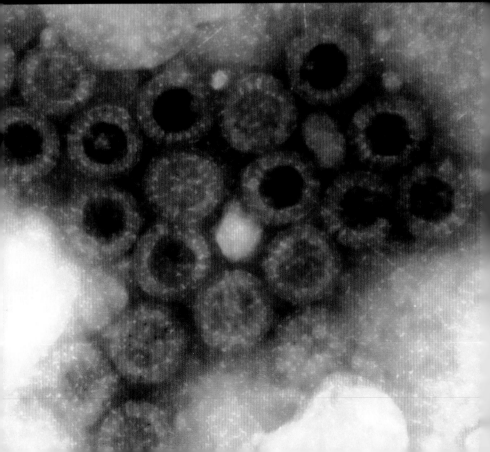

RNA VIRUSES: ASTROVIRUS

astrovirus/diarrheal disease

■ **ORIGIN**

Some astroviruses cause diarrhea in other animals, but these are not the same as human astroviruses.

■ **FIRST IDENTIFIED**

Detected in 1975 and cultured in 1981.

■ **TRANSMISSION**

Feco-orally, directly or indirectly in water or shellfish.

■ **PORTAL OF ENTRY**

Mouth.

■ **INCUBATION PERIOD**

2–3 days.

Astrovirus is a small, round virus (about 30 nm) which on electron microscopic examination appears to have a star stamped on its surface (*astron* is Greek for star). It has a worldwide distribution, is the second or third most common cause of diarrheal disease in children, and can cause epidemics of disease in adults and children alike.

To date, eight different "serotypes" have been detected. Of these, serotype-3 seems to have the greatest ability to cause epidemics of gastro-enteritis and produces the most severe disease. Large numbers of virus particles (up to one million million) are excreted during acute infection. The infective dose is unknown but is thought to be low. In temperate countries, infections peak in the winter months (usually before rotavirus).

CLINICAL FEATURES

After a short incubation period (two to three days), patients present with nausea, vomiting, and acute watery diarrhea, but in general the disease is less severe than that caused by rotavirus. On average it lasts six to seven days and resolves completely.

TREATMENT

No specific antiviral therapy available. Patients must be assessed for degree of dehydration and rehydrated with oral rehydration solution or intravenously, as appropriate.

PREVENTION

No vaccine is available, nor is one likely to be so in the near future.

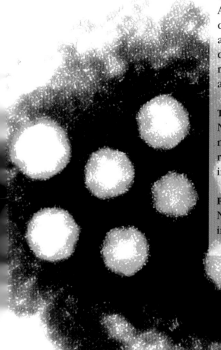

RNA VIRUSES: CALICIVIRUSES

norovirus/gastroenteritis

This small, round virus (27–40 nm) was the first virus to be associated with diarrheal disease. Since then a number of related viruses have been classified as noroviruses, within the caliciviruses family.

Noroviruses are a significant cause of diarrheal disease (particularly vomiting) in children and adults, second only to rotaviruses in importance and frequency. They are excreted in feces, saliva, and vomitus, and spread very easily, especially where people are in close contact. They can also be transmitted in shellfish, which concentrate the virus from sewage outlets.

CLINICAL FEATURES
Incubation can take as little as 12 to 24 hours, and is followed by aversion to food (anorexia), nausea, profound vomiting (10 to 15 times per day), then abdominal pain and diarrhea. The disease usually resolves within five to six days.

TREATMENT
Rehydration, either orally or intravenously. No antiviral drugs are available.

PREVENTION
No vaccine is available, nor likely.

QUICK REFERENCE

■ **ORIGIN**
There are noroviruses of cattle and other animals but these are not closely related to human strains, which have probably evolved with humans.

■ **FIRST IDENTIFIED**
In 1968, after an outbreak in a school in the U.S. town of Norwalk. The virus was first detected by electron microscopy in 1972. It has never been cultured.

■ **TRANSMISSION**
Feco-orally or through saliva or vomiting. Transmits easily and can also be spread by food or in drinking water.

■ **PORTAL OF ENTRY**
Mouth.

■ **INCUBATION PERIOD**
24–72 hours.

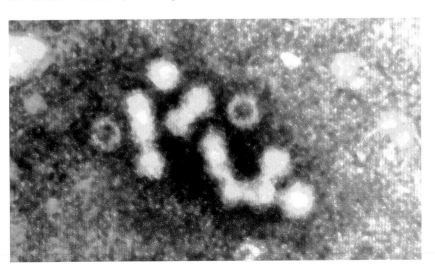

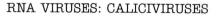

RNA VIRUSES: CALICIVIRUSES
sapovirus/diarrheal disease

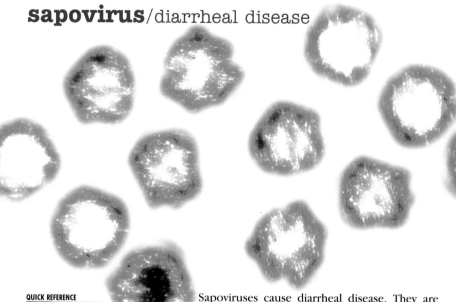

QUICK REFERENCE

■ **ORIGIN**
There are some animal caliciviruses, but these are different from sapovirus. The viruses probably evolved with humans.

■ **FIRST IDENTIFIED**
By electron microscopic examination of feces of children with diarrhea in 1978. It has never been grown in artificial culture.

■ **TRANSMISSION**
Feco-orally.

■ **PORTAL OF ENTRY**
Mouth.

■ **INCUBATION PERIOD**
2–3 days.

Sapoviruses cause diarrheal disease. They are named after Sapporo virus, the first virus of this subfamily to be described. Like norovirus, sapovirus is one of the calicivirus family of viruses. Under electron microscopy, the characteristic "Star of David" shape of caliciviruses makes them appear as if they have cup-shaped depressions in their surface (*calyx* is Greek for a cup).

Sapoviruses are small, unenveloped RNA viruses (27–30 nm). Unlike the noroviruses, they are found predominantly in cases of childhood gastroenteritis and in only a minority (less than 2 percent) of cases of severe gastroenteritis. They do not appear to be associated with foodborne outbreaks.

CLINICAL FEATURES
The incubation period is around two to three days. The virus usually causes a mild acute watery diarrhea.

TREATMENT
The main treatment is assessing the degree of, and appropriately correcting, dehydration. There is no specific antiviral treatment.

PREVENTION
No vaccine is available, nor likely.

RNA VIRUSES: CALICIVIRUSES
hepatitis E/infective hepatitis

Not all cases of infective hepatitis (transmitted feco-orally) are due to the hepatovirus (see page 23). In 1983, electron microscopy detected a further virus, now known as hepatitis E, in feces. The virus is small (27–32 nm) and resembles members of the calicivirus family. In Europe and North America, it is brought in by travelers returning from regions where there is endemic and epidemic disease. Epidemics are usually related to drinking water contaminated by human excreta. Clinically apparent hepatitis generally occurs in people aged 25 to 40. Infection in children is usually unapparent. Most infections in endemic areas are in those over 16 years old, whereas most cases of hepatitis A are in those under five.

CLINICAL FEATURES
After an incubation period of around 40 days, patients present with a self-limiting illness consisting of fever, malaise, aversion to food, and abdominal discomfort, followed rapidly by jaundice. The disease is very similar to that caused by hepatovirus, the only major difference being its impact on women who become infected during the last trimester (third) of pregnancy, where mortality rates rise to between 15 percent and 20 percent for mothers and their babies.

TREATMENT
No specific treatment, only supportive therapy.

PREVENTION
No vaccine is available.

QUICK REFERENCE

■ ORIGIN
Some consider that it is zoonotic (transferred from an animal reservoir).

■ FIRST IDENTIFIED
First recognized as an infection in the 1950s. Virus described in 1983 but not yet grown artificially.

■ TRANSMISSION
Spread feco-orally but in epidemics, drinking water contaminated with human excreta is the major route.

■ INCUBATION PERIOD
Around 40 days.

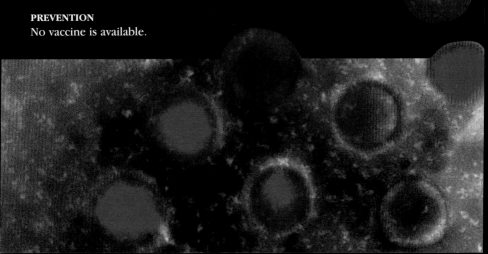

RNA VIRUSES: MYXOVIRUSES

influenza viruses A, B, & C/influenza

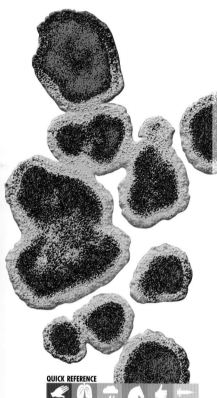

This disease was first described in the 16th century in Italy, who thought it was due to the influence (*influenza* in Italian) of the moon and the stars. However, there is no record of a viral cause (filterable agent) until that which emerged for chicken flu in 1901.

The influenza viruses are highly mutable. On their exterior surface, they have two types of spikes: HA (which attaches the virus to host cells) and NA (which cuts a newly synthesized virus free from the host cell). These represent the major antigens (molecules recognized by the immune system) and antibodies to these provide immunity to infection.

FLU PANDEMICS

Influenza A is the most important pathogen in the family, causing often severe influenza epidemics every winter. And when a brand new strain emerges, the whole world is susceptible, resulting in a pandemic (*pan* is the Greek for "all"). The first major pandemic, known as "Spanish flu," occurred in 1918–19. It spread worldwide and affected all age groups, killing more people (20 million) than were killed during World War I. This virus continued to circulate causing epidemics of influenza during which only a proportion of the population became ill.

In 1957, a new strain emerged, leading to the "Asian flu" pandemic. Next came "Hong Kong flu" in 1968. We are now overdue for a new pandemic. In 1997, an avian strain caused a small outbreak of infection in Hong Kong in which six of the 18 infected patients died. However, the strain did not easily transmit from person to person and therefore did not cause serious concern with regard to the threat of a pandemic. Nevertheless, this virus is still around and will require only a few mutations before it becomes problematic.

The question is, how does the virus keep changing in order to keep coming back in new guises? Influenza A virus changes in two major ways, called antigenic drift and antigenic shift.

QUICK REFERENCE

■ **ORIGIN**
Continually evolving from bird and animal viruses.

■ **FIRST IDENTIFIED**
Disease described as long ago as the 16th century. Influenza A was first grown in 1933, influenza B in 1940, and influenza C in 1947.

■ **TRANSMISSION**
Airborne or via respiratory secretion on hands.

■ **PORTALS OF ENTRY**
Mouth and upper respiratory tract.

■ **INCUBATION PERIOD**
2–5 days.

Antigenic drift is the slow changing of viruses that accounts for the yearly epidemics when a proportion of the population is infected. Antigenic shift is responsible for the major changes in viruses, such as that which caused Spanish flu.

CLINICAL FEATURES

Influenza virus infection can range from being totally asymptomatic to severe life-threatening pneumonia. In general, influenza A produces the most severe disease followed by B and then C. It should also be remembered that a number of other viruses can produce similar disease manifestations. Patients present with fever, chills, headache, muscle, and joint pains, sore throat, dry cough, and aversion to food (anorexia). Their temperature usually peaks over the first 24 hours (at 104–106°F/40–41°C) but persists for up to five days. In most cases, the illness resolves but, particularly in the elderly and those with damaged hearts or lungs, can lead to severe chest infection.

TREATMENT

Usually, treatment is not necessary. However, in severe cases, there are drugs that can be used prophylactically (amantidine) after contact with a case of flu, or for treatment (zaminovir).

PREVENTION

A number of vaccines are available. The most frequently used vaccine consists of purified HA and NA spikes (subunit vaccine), which is given by injection. A new vaccine is made each year, incorporating the influenza A and B subtypes expected to cause disease the next season. It is recommended for those over 60 and anyone with chronic heart or lung disease. Revaccination is required each year.

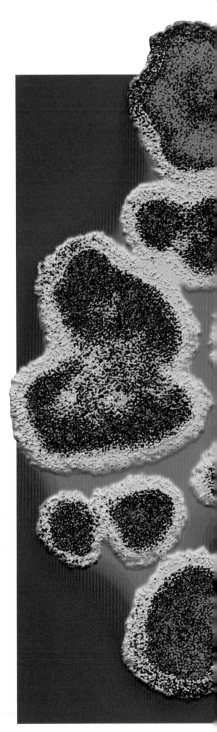

RNA VIRUSES: MYXOVIRUSES
parainfluenza viruses/
respiratory disease group

These are enveloped RNA viruses related to influenza viruses (*para* implies "related to" in Greek), but their genome is one long strand of RNA rather than segmented, which means they do not show antigenic shift (see pages 30–31). There are five human parainfluenza viruses (HPIVs): 1, 2, 3, 4a, and 4b. Viruses 1 and 3 are related to each other, but 2, 4a, and 4b are more closely related to the mumps virus.

All HPIVs infect the respiratory tract causing a variety of different respiratory diseases. HPIV1 causes outbreaks of disease, usually in the fall, every second year, and HPIV2 causes outbreaks, again in the fall, every year. HPIV3 causes infection throughout the year with peaks in the spring. The patterns of HPIV4a and 4b are not well studied. The viruses are excreted in respiratory secretions and transmitted from person to person via large airborne droplets ("coughs and sneezes spread diseases"), on hands, or even on inanimate objects such as toys or handkerchiefs.

CLINICAL FEATURES
After a short incubation period, patients present with fever, malaise, coryza (catarrhal inflammation of the mucous membrane in the nose), and cough, rather like a common cold. What happens next depends on which virus is causing the infection, the age of the patient, and the individual's immune system. HPIV1 generally causes croup (infection and inflammation of the larynx and trachea). HPIV2 usually causes only a common cold. HPIV3 is second only to respiratory syncytial virus as a cause of pneumonia or bronchiolitis, and HPIV4a and 4b tend to cause only mild illness.

TREATMENT
No safe and effective treatment is available.

PREVENTION
No vaccine is available currently, but live attenuated vaccines are being developed.

QUICK REFERENCE

■ **ORIGIN**
Probably evolved with humans.

■ **FIRST IDENTIFIED**
HPIV1, 2, and 3 cultured in 1956.

■ **TRANSMISSION**
Airborne via coughing and sneezing, or on hands or fomites carrying dried respiratory secretions.

■ **PORTALS OF ENTRY**
Mouth or upper respiratory tract.

■ **INCUBATION PERIOD**
1–5 days.

RNA VIRUSES: MYXOVIRUSES
mumps virus/mumps

Mumps used to be a common acute infectious disease affecting children in Europe and North America. However, since the initiation of vaccination using the mumps, measles, and rubella (MMR) live attenuated vaccine, incidences of the disease have decreased markedly. Prior to MMR, up to 80 percent of children in urban areas experienced infection. In smaller, more remote communities, outbreaks of mumps occurred every two to seven years. In temperate countries, infection peaks in the winter and spring.

CLINICAL FEATURES
Patients complain of fever, muscle pains, and malaise. The salivary glands at the side of the face (parotid glands) swell up, usually bilaterally (on both sides of the face), which takes around 10 days to resolve, though in about a third of childhood infections this does not occur. Patients are infective (excreting virus) for between five and six days prior to disease and for five to six days after symptoms begin. About 5 percent of patients may develop meningitis.

If males over the age of puberty are infected, around a quarter develop orchitis (inflammation of the testicles). This is bilateral in one in five sufferers. Occasionally, post-pubertal females develop oophoritis (infection of the ovaries).

TREATMENT
No specific antiviral therapy is available.

PREVENTION
The live attenuated vaccine incorporated into MMR is given to all children aged 12–18 months. It is very effective, though immunity declines with age.

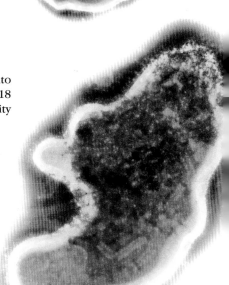

RNA VIRUSES: MYXOVIRUSES

measles virus/measles

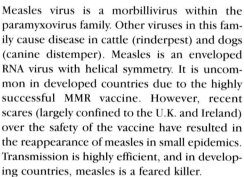

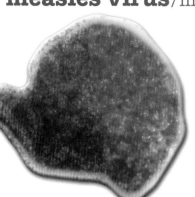

■ **ORIGIN**
It is thought that it first emerged as long ago as 2500 B.C. from the cattle disease, rinderpest. Its name is derived either from *misellus* or *mazer*, which are the Latin and Teutonic words meaning miserable.

■ **FIRST IDENTIFIED**
Virus was first grown in monkeys in 1905 and in artificial culture in 1954.

■ **TRANSMISSION**
Via respiratory droplets or on hands.

■ **PORTALS OF ENTRY**
Mouth, upper respiratory tract, or conjunctivae.

■ **INCUBATION PERIOD**
9–11 days.

Measles virus is a morbillivirus within the paramyxovirus family. Other viruses in this family cause disease in cattle (rinderpest) and dogs (canine distemper). Measles is an enveloped RNA virus with helical symmetry. It is uncommon in developed countries due to the highly successful MMR vaccine. However, recent scares (largely confined to the U.K. and Ireland) over the safety of the vaccine have resulted in the reappearance of measles in small epidemics. Transmission is highly efficient, and in developing countries, measles is a feared killer.

CLINICAL FEATURES
Patients develop fever, coryza (catarrh), and conjunctivitis, and small white/yellow spots (Koplik's spots) appear on the internal surface of the cheeks. After a day or so, the measles rash, which consists of red flat or raised spots that often merge together, appears first on the hairline on the neck then rapidly spreads to the face and trunk, reaching the extremities over the next three days. Possible complications include pneumonia, middle-ear infections, encephalitis and, rarely, subacute sclerosing panencephalitis (a progressive fatal loss of brain cells).

TREATMENT
No antiviral therapy is available.

PREVENTION
The MMR vaccine is highly effective, and concerns that it might cause autism unfounded.

RNA VIRUSES: MYXOVIRUSES

human metapneumovirus/
respiratory tract infection

In 2001, virologists in Holland "discovered" human metapneumovirus (hMPV) while examining the causes of acute respiratory tract infection in infants. However, it is now known that this worldwide virus is not in fact new. It is responsible for a proportion of the respiratory tract infections that occur in temperate countries each year, and has a seasonal distribution similar to respiratory syncytial virus (RSV—see pages 36–37) with peaks each fall and winter. It is transmitted in exactly the same way as RSV and some infants can be infected with both RSV and hMPV. It is suggested that this coinfection might result in more severe disease.

CLINICAL FEATURES
Clinical characteristics in infants are similar to those associated with RSV, though usually less severe and responsible for far fewer cases of lower respiratory tract infection. However, there is some evidence that hMPV may cause more wheezing than RSV.

TREATMENT
No antiviral drugs are available.

PREVENTION
No vaccine is available.

QUICK REFERENCE

■ **ORIGIN**
There is a metapneumovirus of turkeys but it is not able to infect humans, and hMPV does not infect birds.

■ **FIRST IDENTIFIED**
In 2001, in the Netherlands.

■ **TRANSMISSION**
By respiratory droplets or dried (or moist) respiratory secretions on hands.

■ **PORTAL OF ENTRY**
Upper respiratory tract.

■ **INCUBATION PERIOD**
1–5 days.

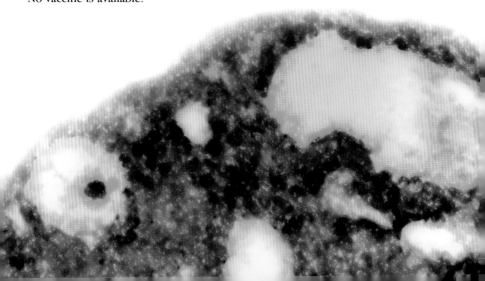

RNA VIRUSES: MYXOVIRUSES

respiratory syncytial virus/
bronchiolitis

Respiratory tract infections are the most common infective cause of death worldwide and the third most common cause of death overall. Deaths due to respiratory tract infection tend to occur in the very young or the elderly.

Infections are usually divided into upper respiratory tract infections (URIs) and lower respiratory tract infections (LRIs). It is estimated that each child experiences between five and 10 URIs each year up to the age of five. This incidence rate is the same in both developed and developing countries. However, URI can progress to LRI, which can result in death, particularly in developing countries as a result of factors such as malnutrition, vitamin A deficiency, and smoke pollution.

The main cause of URI and LRI in infants is respiratory syncytial virus (RSV), a pneumovirus of the paramyxovirus family. RSV infects the full length of the respiratory tract. It is transmitted via airborne respiratory secretions or, more importantly, by the transfer of respiratory secretions on hands. By the age of three, every child has at some time been infected with RSV, and it is likely that each of us will become infected

every year throughout our lifetimes. RSV is responsible for 25 hospitalizations per 1,000 infants annually in the U.K., 3 percent of which are so severe that ventilatory support in an intensive-care unit is necessary. In temperate countries, RSV infections peak in the fall and winter, and in tropical countries, in the rainy seasons.

CLINICAL FEATURES

The incubation period is short, at between one and five days, and manifestations vary from a common-cold-like illness to bronchiolitis and pneumonia. In children over five and young adults it usually causes URI accompanied by a sore throat. In infants, it can cause bronchiolitis in which the bronchioles (the fine tubes of the lungs) become inflamed and narrow so that although air can be inhaled, exhalation is slow and less complete (not unlike an asthma attack). The child's lungs hyperinflate and because they cannot exhale they do not remove the build-up of carbon dioxide nor take in sufficient oxygen. Eventually this can lead to heart failure as the heart beats faster and faster to try to oxygenate the body.

TREATMENT

The antiviral drug ribavirin works in vitro but is less effective in clinical trials, and the antibody monoclonal is used prophylactically (as a method of prevention), and sometimes therapeutically in (usually premature) babies with damaged lungs.

PREVENTION

As yet there is no safe and effective vaccine, although a number of candidate vaccines look hopeful. A previous formalin-inactivated RSV vaccine produced more severe disease in recipients when they became infected and has therefore been withdrawn.

QUICK REFERENCE

■ ORIGIN
There are respiratory syncytial viruses of cattle and other animals but these are distinct from human RSV.

■ FIRST IDENTIFIED
First suggestion of a viral etiology in 1939, and virus first isolated (from a chimpanzee and from the zoo-keeper who also became infected) in 1956.

■ TRANSMISSION
Airborne transmission of respiratory secretions or, more importantly, on hands.

■ PORTAL OF ENTRY
Upper respiratory tract.

■ INCUBATION PERIOD
1–5 days, with patients excreting virus for up to 3 weeks after the onset of disease.

RNA VIRUSES: MYXOVIRUSES

nipah and hendra viruses/
encephalitis, LRI

■ **ORIGIN**
The viruses naturally infect fruit bats and cause severe disease when transmitted to humans and other animals.

■ **FIRST IDENTIFIED**
Hendra was first described in 1994 and Nipah in 1998–99.

■ **TRANSMISSION**
Mostly by respiratory secretions and inhalation.

■ **PORTAL OF ENTRY**
Upper respiratory tract.

■ **INCUBATION PERIOD**
4 days to 2 months (but usually less than 2 weeks) for Nipah virus.

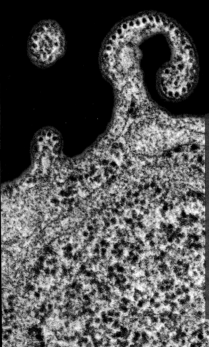

These two paramyxoviruses are newly emerged causes of human infections, which are zoonotic (transferred from animals to humans). Hendra is named after the town in Queensland, Australia, where in 1994 the virus caused an outbreak of severe respiratory tract infection in horses and two humans, who were in close contact with the horses. Eventually it was discovered that the reservoir host was the fruit bat, which is persistently infected and excretes the Hendra virus in its urine and saliva.

The Nipah virus emerged in Malaysia in 1998–99 where it caused a large outbreak of encephalitis, killing more than 100 of those infected. Many of the survivors are now developing a disease similar to the subacute sclerosing panencephalitis that can follow measles. The outbreak occurred after large pig farms were set up in previously forested jungle areas where the domestic pigs came into contact with fruit bats, who dropped partially chewed fruit contaminated with the virus into the pig enclosures. The pigs subsequently became infected, coughing out large amounts of the Nipah virus, infecting pig farmers, vets, and abattoir workers.

CLINICAL FEATURES
Hendra causes a severe lower respiratory tract infection in humans, and Nipah causes pneumonia and encephalitis in pigs. Humans develop encephalitis with progression to coma and death in up to 50 percent of cases.

TREATMENT
Ribavirin has shown some promise in the treatment of Nipah virus infections.

PREVENTION
There is no vaccine. The Nipah outbreak was halted by culling over a million pigs. However, fruit bats are still excreting both viruses and human infection could recur.

RNA VIRUSES: FLAVIVIRUSES

yellow fever virus/
hepatitis, hemorrhagic fever

Yellow fever virus is part of the large group of arboviruses (or arthropod-borne viruses) transmitted by insects. Blood-feeding insects such as mosquitoes, ticks, and sandflies take infected blood from humans and other animals. The virus grows inside the insect, which then injects it into whichever human is the insect's next target. This is termed biological transmission, whereby the virus multiplies in the vector insect, as opposed to mechanical transmission in which there is no growth of the virus.

It is now clear that there are a number of different virus families that are arboviruses, including flaviviruses, the most major of which is yellow fever (*flavus* is Latin for yellow). The first recorded outbreak of yellow fever was in Mexico in 1648. Also called Yellow Jack, the virus was a major problem for crews of British ships visiting Africa, South and Central America, and the West Indies, and even caused disease in the U.S.A. as far north as New York. It is transmitted to humans by the aedes mosquito, either from other humans (the urban cycle) or wild monkeys (the sylvatic cycle), the latter being less severe.

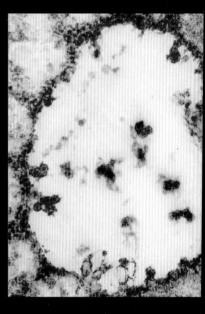

CLINICAL FEATURES
Illness presents suddenly with chills, fever, and headache. This is followed by muscle pain (myalgia), including severe back pain. After three to four days, moderately ill patients will begin to recover. However, in some patients, the temperature falls then returns with signs of shock, jaundice, and unstoppable nosebleeding and bleeding of the gums, which results in black vomit. Fatality rates are as high as 50 percent.

TREATMENT
No specific antiviral therapy is available.

PREVENTION
The live attenuated virus vaccine (17-D) is safe and highly effective. However, limiting the growth of the mosquito vector is also important.

QUICK REFERENCE

■ **ORIGIN**
Probably a monkey virus that has adapted to infect humans.

■ **FIRST IDENTIFIED**
Outbreak in Mexico in 1648. The transmission cycle involving mosquitoes was described by Finlay in Cuba and proved by Reed and Gorgas in 1918. The virus was first cultured artificially in 1937 in mouse embryo cells.

■ **TRANSMISSION**
From humans or monkeys via the aedes mosquito.

■ **PORTAL OF ENTRY**
Injected through the skin to the bloodstream by the feeding mosquito.

■ **INCUBATION PERIOD**
3–6 days.

RNA VIRUSES: FLAVIVIRUSES
dengue virus/viral hemorrhagic fever

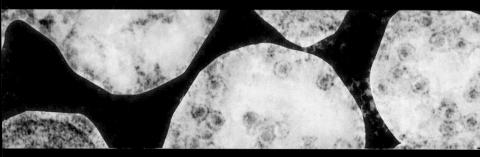

QUICK REFERENCE

■ **ORIGIN**
First probable description of dengue was of an outbreak in Philadelphia, U.S.A., in 1780. The virus has likely evolved from other, less pathogenic, strains.

■ **FIRST IDENTIFIED**
Transmission via mosquitoes was suggested in 1906 and proven in 1929. The virus was first grown in artificial culture in 1960.

■ **TRANSMISSION**
Various mosquitoes.

■ **PORTAL OF ENTRY**
Injected through the skin and released into the bloodstream.

■ **INCUBATION PERIOD**
2–7 days.

Dengue is transmitted by mosquitoes. The reservoir hosts are usually other infected humans but a monkey sylvatic cycle may also exist. Epidemics occur after periods of rainfall, which create stagnant water where the mosquitoes can breed. The mosquitoes feed on dengue-infected humans and the virus is established in their salivary glands, ready to inject during the insects' next feed.

Mosquitoes do not transmit the virus at temperatures below 79°F (26°C), but global warming is resulting in the emergence of dengue in previously unaffected areas.

CLINICAL FEATURES
Young children tend to present a fever with a rash, and those under 15 an asymptomatic infection. Classic dengue occurs in adolescents and adults. It has an abrupt onset with fever, severe headaches, joint and muscle pain, and a confluent flat or raised red rash. Recovery can be prolonged and includes depression.

Dengue hemorrhagic fever tends to occur in hyperendemic regions, usually in children under 15. There is bleeding into skin and mucous membranes, shock, and liver enlargement and tenderness in up to 40 percent of cases. Mortality rates of 20 percent are not uncommon, and can be as high as 40 percent where there is shock.

TREATMENT
Treatment is supportive.

PREVENTION
No safe and effective vaccine is available.

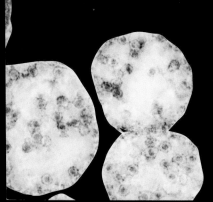

RNA VIRUSES: FLAVIVIRUSES

japanese encephalitis virus/
meningitis and encephalitis

Japanese encephalitis virus (JEV) is the most significant neutrotropic flavivirus. There are an estimated 50,000 cases per year with as many as 15,000 deaths. It is found in Southeast Asia but is on the move, with cases now being reported in India, much of southern China, the Pacific Rim, and northern Australia.

JEV is naturally transmitted between birds and animals by culicine mosquitoes. Humans become infected when they accidentally stray between the two hosts. The pig is one of the main reservoir hosts, and is infected asymptomatically, except when pregnant (this results in spontaneous abortion).

Most children in rural areas of Southeast Asia appear to be infected with JEV, but only one in 300 actually develops the disease. Epidemics of JEV occur especially when mosquito populations expand after the rains. The virus can also be transmitted down through the generations (transovarial spread). JEV is a risk for foreign travelers to endemic areas, especially if they are spending more than two weeks in rural areas.

CLINICAL FEATURES
Most infections result in a febrile illness with no specific pattern. In some, the febrile illness progresses to coma and fits (signs of encephalitis). It can also present with neck stiffness (meningitis) or even a polio-like flaccid paralysis. Some 20 percent to 30 percent of patients with neurological manifestations of disease die.

TREATMENT
No specific antiviral therapy is available.

PREVENTION
The available formalin-killed whole virus vaccine is effective and safe, but relatively expensive, which means it cannot be used widely in endemic areas. A cheaper and possibly more effective Chinese live attenuated virus vaccine is also in existence, but is not yet licensed for use in Europe and the U.S.A.

QUICK REFERENCE

■ **ORIGIN**
Seems to have evolved from a non-pathogenic precursor in Indonesia around 1850.

■ **FIRST IDENTIFIED**
The first cases described in Japan in 1871. Virus first isolated in 1935.

■ **TRANSMISSION**
By culicine mosquitoes from birds or animals (particularly pigs).

■ **PORTAL OF ENTRY**
Injection through the skin and into the bloodstream by feeding mosquitoes.

■ **INCUBATION PERIOD**
5–14 days. Progression from infection to death can be as quick as 10 days.

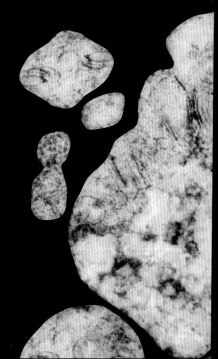

RNA VIRUSES: FLAVIVIRUSES

west nile virus/west nile encephalitis

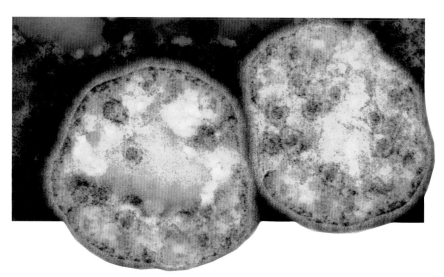

QUICK REFERENCE

■ **ORIGIN**
A virus of birds accidentally transmitted to humans.

■ **FIRST IDENTIFIED**
In Uganda in 1937. Crossed to North and Central America in 1999.

■ **TRANSMISSION**
From birds to humans by culicine mosquitoes.

■ **PORTAL OF ENTRY**
Through the skin to the bloodstream.

■ **INCUBATION PERIOD**
3–6 days.

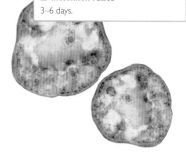

This flavivirus was, until recently, thought to be relatively benign. It was found across Africa, much of Asia, and even southern Europe, but has now lived up to its name and spread much further west of the Nile. It has become endemic in the U.S.A. and Canada (probably via migrating birds) and has also spread south to Mexico. It is transmitted between birds by culicine mosquitoes, and humans are infected by confused mosquitoes. There is no onward transmission from humans.

CLINICAL FEATURES
Large outbreaks in South Africa and Israel in the 1950s and 1970s caused a febrile illness with joint pains and a rash similar to the dengue rash. However, in a large outbreak in Romania in 1996, more than 600 patients suffered neurological problems. The elderly are much more likely to get encephalitis with convulsions and coma.

TREATMENT
No specific antiviral therapy, but some patients treated with ribavirin have recovered.

PREVENTION
No vaccine is available. Prevention relies on mosquito control, including larvicidal sprays.

RNA VIRUSES: HEPACIVIRUS

hepatitis C virus/hepatitis

Hepatitis C virus (now called hepacivirus) is not transmitted by insects, but by injecting drug-abusers who share needles and syringes (as many as 80 percent may now be infected), via blood transfusions or blood products, organ transplantation, sexual intercourse, and transfer from mother to baby. It is highly mutable and maintained in the population by carriers, of which it is estimated there are 170 million worldwide, though prevalence varies geographically.

CLINICAL FEATURES

Patients may develop classic signs of hepatitis (fever, jaundice, and abdominal discomfort), but most are infected asymptomatically. Over 80 percent of patients fail to eliminate the virus and become chronic carriers. Most live a normal life but some develop chronic liver damage and failure (cirrhosis), or cancer of the liver.

TREATMENT

A combination of ribavirin and interferon-alpha slows down virus replication and the progression of liver disease.

PREVENTION

No vaccine is available, nor likely in the near future. Prevention depends on avoiding the sharing of needles and syringes and the screening of blood products and organ donors.

QUICK REFERENCE

■ **ORIGIN**
Unknown.

■ **FIRST IDENTIFIED**
Suspicions arose in the 1960s, but the virus was not detected until 1989.

■ **TRANSMISSION**
Via the blood.

■ **PORTAL OF ENTRY**
To the bloodstream and then the liver.

■ **INCUBATION PERIOD**
6–9 weeks.

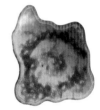

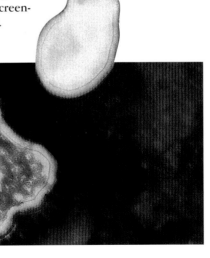

RNA VIRUSES: ALPHAVIRUSES

eastern and western equine encephalitis/encephalitis

QUICK REFERENCE

■ **ORIGIN**

Viruses that naturally infect birds are accidentally transmitted to humans, via horses, by mosquitoes.

■ **FIRST IDENTIFIED**

WEV was first isolated from horses in 1930 and from humans in 1938. EEV was first isolated from a horse in 1933 and from humans in 1938.

■ **TRANSMISSION**

Usually by mosquitoes from horses. Birds are the ultimate reservoir hosts.

■ **PORTAL OF ENTRY**

Injected through the skin by the mosquito's proboscis into the bloodstream.

■ **INCUBATION PERIOD**

5–10 days.

These viruses cause human encephalitis in the eastern and western states of the U.S.A. Eastern equine encephalitis (EEV) is transmitted between birds by a variety of mosquitoes in swampy and forested areas. It has also been detected in Canada, Central America, and the Caribbean. There are major outbreaks of disease in horses, but fortunately, human infection is rare; EEV has a 50 percent mortality rate.

Western equine encephalitis (WEV) is transmitted between passerine birds by culicine mosquitoes. Human infections follow those in horses and are also transmitted by the same mosquitoes. There is no onward transmission.

CLINICAL FEATURES

Illness begins with a headache and a high fever. EEV produces more severe disease, but in many cases, and especially with WEV, the illness progresses no further. However, the disease can progress to coma and death.

TREATMENT

No antiviral therapy is available.

PREVENTION

No human vaccine, but a horse vaccine reduces transmission to humans. Mosquito control is another preventative intervention.

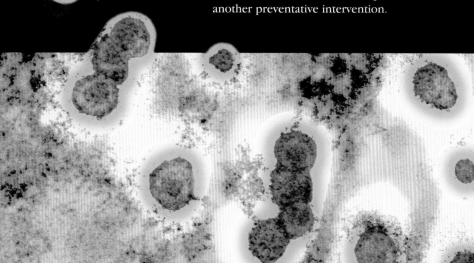

RNA VIRUSES: ALPHAVIRUSES

o'nyong-nyong virus/
breakbone fever

The native name of this alphavirus means "jointbreaker" in the language of the Ugandan Acholi tribe. Although originally isolated in Uganda, there have also been cases of disease in the Central African Republic, Kenya, Malawi, Mozambique, Senegal, and Tanzania. It is transmitted to, and probably between, humans by anopheline mosquitoes including *Anopheles funestus* and *An. gambiae*. The latter is a particularly avid biter of humans. No animal reservoir of the virus has been detected.

Infections appear in waves. Between 1959 and 1962, two million people were infected in East Africa, and in some areas, 70 percent of the population was infected. In 1996 and 1997, an outbreak in Uganda had attack rates of between 29 percent and 35 percent in affected areas.

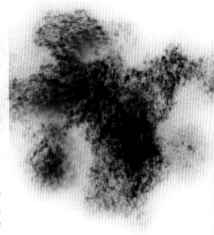

CLINICAL FEATURES

Patients present with fever followed by a sudden onset of muscle and joint pains, pain around the eyes, and a rash. Some also experience chills and nosebleeds prior to the fever. The lymph glands throughout the body, and especially around the neck, become enlarged (lymphadenopathy). The infection can take weeks to resolve but it does not result in death nor any long-lasting damage. Approximately half of those infected remain asymptomatic.

TREATMENT

No antiviral treatment is available.

PREVENTION

No vaccine is available. Avoiding mosquito bites is the only way of preventing infection.

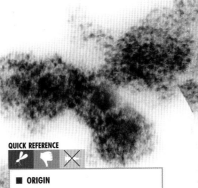

QUICK REFERENCE

■ **ORIGIN**
Unknown.

■ **FIRST IDENTIFIED**
Virus isolated in 1959.

■ **TRANSMISSION**
Probably human-to-human, by anopheline mosquitoes.

■ **PORTAL OF ENTRY**
Injected by the mosquito's proboscis through the skin, into the bloodstream.

■ **INCUBATION PERIOD**
8 days or longer.

RNA VIRUSES: ALPHAVIRUSES

rubella virus/german measles

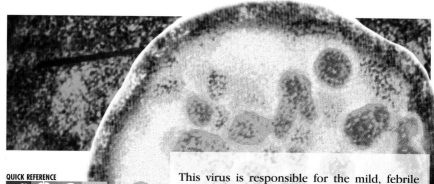

QUICK REFERENCE

■ **ORIGIN**
Probably evolved with humans.

■ **FIRST IDENTIFIED**
Disease defined by two German physicians in the 1800s. Rubella virus first isolated in 1962.

■ **TRANSMISSION**
Person-to-person by respiratory secretions, either airborne or on hands. Virus begins to be excreted about 7 days before the rash appears and for 21 days afterwards. Congenitally infected babies excrete the virus for months or even years.

■ **PORTALS OF ENTRY**
Upper respiratory tract and mouth.

■ **INCUBATION PERIOD**
2–3 weeks.

This virus is responsible for the mild, febrile infectious childhood disease known as rubella, or German measles. However, in nonimmune pregnant women, rubella is able to cross the placenta to infect the fetus, resulting in intrauterine death, mental retardation, cardiac defects, deafness, and cataracts. Humans are the only hosts.

CLINICAL FEATURES
Patients develop a fever and a red pinpoint rash beginning on the face and spreading to the rest of the body. Lymph nodes become enlarged, particularly on the back of the neck. Encephalitis occurs in one in 10,000 cases and adults tend to get joint pains. If a woman is infected for the first time during the first 10 weeks of pregnancy, there is an 80 percent risk of death of the fetus. If infection occurs after 10 weeks, babies are born with congenital defects.

TREATMENT
No specific antiviral therapy is available. Pooled immoglobulin containing antibodies to the rubella virus is given to pregnant women, as this may decrease the extent of fetal damage.

PREVENTION
The live attenuated rubella virus vaccine is given as part of the MMR in 12- to 15-month-old babies, with a booster at age 12 to 14 years. It provides excellent protection against rubella, especially congenital rubella.

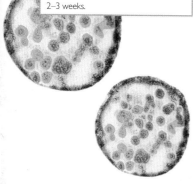

RNA VIRUSES: ARENAVIRUSES

lassa fever virus/
viral hemorrhagic fever

Lassa fever virus (LFV) is zoonotic (transferred from animals to humans) and person-to-person spread also occurs. The reservoir host is the multimammate rat, found particularly in West Africa. The virus is excreted in large amounts through the rat's urine, and humans become infected by inhalation of the virus in the dried urine.

Areas where infection is common include Nigeria, Liberia, Sierra Leone, Côte d'Ivoire, and Guinea. Previously feared as a cause of fatal viral hemorrhagic fever, it is now clear that this widespread infection can be mild. However, it is estimated that LFV causes 5,000 deaths each year in West Africa.

CLINICAL FEATURES
LFV can cause a mild febrile illness, beginning with the gradual onset of fever, headache, and muscle and joint pains. The fever increases over the next few days and is accompanied by abdominal and chest pains. Patients become increasingly lethargic, develop edema (water retention in the tissue) of the face and neck, and bleeding from gums, nose, intestine, and vagina. The mortality rate is around 40 percent.

TREATMENT
If given early, ribavirin is beneficial.

PREVENTION
No vaccine is available, but ribavirin can help (see above).

QUICK REFERENCE

■ **ORIGIN**
A harmless virus of the multimammate rat which causes severe disease when accidentally infecting humans.

■ **FIRST IDENTIFIED**
First recorded case was that of a nurse in Nigeria in 1969. The virus was isolated in 1970.

■ **TRANSMISSION**
Via inhalation of dried urine from the multimammate rat. Subsequent transmission is from contact with the blood and bodily secretions of patients.

■ **PORTALS OF ENTRY**
Mouth and upper respiratory tract.

■ **INCUBATION PERIOD**
1–3 weeks.

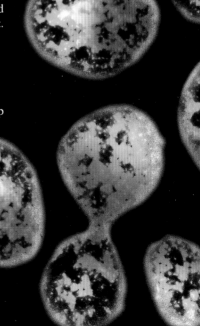

RNA VIRUSES: ARENAVIRUSES

junin, machupo, guanarito, and sabia viruses/
south american hemorrhagic fever

■ ORIGIN
The disease is transmitted to humans from rodents, which are unaffected by carrying it.

■ FIRST IDENTIFIED
Junin virus first isolated in 1958, machupo in 1965, guanarito in 1991, and sabia in 1994.

■ TRANSMISSION
By inhalation or by inoculation of rodent urine or saliva into cuts and grazes.

■ PORTALS OF ENTRY
Through cuts and grazes or the upper respiratory tract.

■ INCUBATION PERIOD
1–2 weeks.

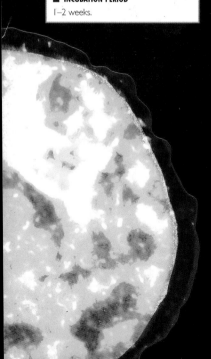

These four viruses are all members of the arenavirus family, each causing hemorrhagic fever in different countries of the New World: junin in Argentina, machupo in Bolivia, guanarito in Venezuela, and sabia in Brazil. Infection is zoonotic (transferred from animals to humans) and the reservoir hosts are small rodents such as the vesper mouse. The rodents are silently and persistently infected with the virus and excrete it in their saliva and urine.

Humans usually become infected by inoculation of rodent urine or saliva into cuts or grazes, or by inhalation of dried urine. Junin epidemics occur every year, with between 100 and 4,000 cases, mostly from April to June (fall in Argentina). Person-to-person spread is mainly between doctors and nurses in hospitals.

CLINICAL FEATURES
Disease begins with a "flu-like" illness. Patients develop back pain, abdominal pain, headaches, and dizziness. Bleeding from the gums or after mild trauma occurs in the first week of illness. During the second week, patients usually begin to improve. However, some develop extensive hemorrhage into the skin (small spots known as petechiae, or larger bruise-like spots called purpura) and mucous membranes. Overall the mortality rate varies between 5 percent and 30 percent.

TREATMENT
Intravenous ribavirin may be beneficial.

PREVENTION
A junin vaccine is proving effective in Argentina. No vaccines available for machupo, guanarita, or sabia. Rodent control is the only other preventative measure available.

RNA VIRUSES: NAIROVIRUS

crimean congo hemorrhagic fever virus/hemorrhagic fever

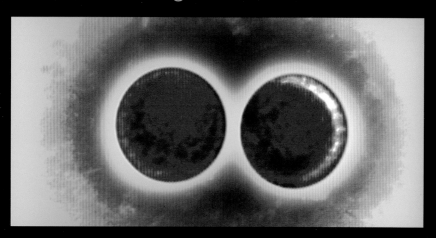

Crimean Congo hemorrhagic fever virus (CCHFV) is a severe illness mainly transmitted via ticks, but which once established, can transmit person-to-person via blood and other bodily secretions. Ticks may also acquire the virus by feeding on small mammals and domestic ruminants, and humans can also become infected by direct contact with the blood of an infected sheep or goat.

CLINICAL FEATURES
Onset is abrupt, with fever, headaches, nausea, muscle pain, and weakness. This lasts for two to three days and is followed by areas of hemorrhage (petechiae and purpura spots) in the skin, bleeding gums and nosebleeds. The petechiae and purpura then coalesce to produce very large areas of bleeding (ecchymoses). Hematemesis (vomiting blood), melena (black stools), and hematuria (blood in urine) are signs of impending death. The mortality rate can be as high as 40 percent.

TREATMENT
Ribavirin may be of some benefit.

PREVENTION
No vaccine is available. Preventing exposure to

QUICK REFERENCE

■ **ORIGIN**
Unknown.

■ **FIRST IDENTIFIED**
In an epidemic in Crimea in 1945. The virus was first isolated in Congo in 1967.

■ **TRANSMISSION**
By tick bite or exposure to blood or bodily secretions of infected animals or humans.

■ **PORTAL OF ENTRY**
Through the skin.

■ **INCUBATION PERIOD**
3–6 days if acquired person-to-person, or 12 days if transmitted by ticks.

RNA VIRUSES: PHLEBOVIRUS

sandfly fever/phlebotomus fever

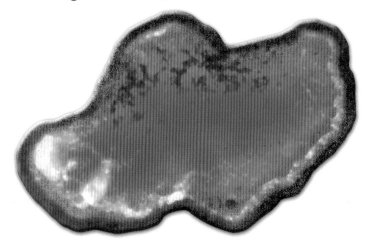

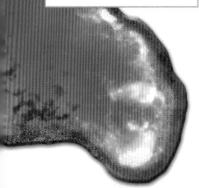

As its name implies, sandfly fever virus is transmitted by the sandfly, and the only reservoir hosts are humans. However, the virus is transmitted transovarially down the generations of sandflies, which breed in moist soil in dark, humid locations, and during the winter as larvae in cracks in the soil, beneath stones, and in animal burrows. They emerge as adults in April and infections occur in the warm summer months when the flies are most abundant.

Sandflies are primarily nocturnal feeders, so being outdoors at night, as well as sleeping in infested areas, are risk factors for infection.

CLINICAL FEATURES
Infection begins with the sudden onset of fever that lasts between two and four days, accompanied by severe headaches, orbital pain, pain in the back and joints, anorexia, and malaise. Recovery is complete and within 5 to 10 days.

TREATMENT
The illness is self-limiting and treatment is not required, but ribavirin has been shown to work in experimental infection.

PREVENTION
No vaccine is available. Prevention is by avoiding sandflies and using insect repellents.

RNA VIRUSES: HANTAVIRUSES

hantaan virus/
hemorrhagic fever with renal syndrome

The Hantaan virus does not have an insect vector but instead has an animal reservoir host—the striped fieldmouse. The virus causes no harm to the host and is excreted in urine and saliva.

Humans become infected via inhalation of dried urine or, less frequently, by inoculation of urine or saliva into cuts and grazes. The virus is named after a river in Korea, and produces the most severe disease. Human infections are found in Korea, Japan, China, and Asian Russia.

CLINICAL FEATURES
The most severe form begins with severe abdominal and back pain, muscle pain, headache, dizziness, and petechiae. Next comes tachycardia (fast heartbeat), low blood-pressure, and altered mental state. The following oliguric phase is where patients pass only small amounts or no urine and suffer severe hemorrhage (almost half of all deaths occur during this phase). The diuretic phase signals the beginning of recovery, though convalescence may take months. The overall mortality rate is between 20 percent and 30 percent.

TREATMENT
Treatment is supportive and includes fluid replacement and cardiac support.

PREVENTION
No vaccine is available, so prevention is by limiting contact with rodent excreta.

QUICK REFERENCE

■ **ORIGIN**
Rodents, who are unaffected by carrying it. The viruses can be fatal to humans.

■ **FIRST IDENTIFIED**
A disease similar to Hantaan was described in Chinese literature over 1,000 years ago. In modern times, it caused severe infections in frontline troops (3,000 cases) in the Korean War. Hantaan virus was first cultured in 1981.

■ **TRANSMISSION**
By inhalation of dried excreta from the rodent host.

■ **PORTAL OF ENTRY**
Upper respiratory tract.

■ **INCUBATION PERIOD**
2–42 days (but usually 2–4 weeks).

RNA VIRUSES: HANTAVIRUSES

sin nombre virus/hantavirus
pulmonary syndrome

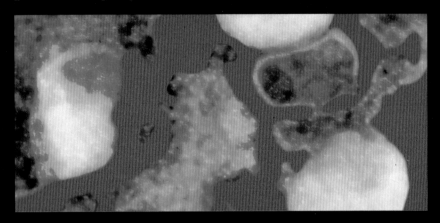

■ ORIGIN
These viruses are harmless to their rodent carriers but when they accidentally infect humans they cause severe disease—a classic zoonosis.

■ FIRST IDENTIFIED
Disease, virus, and reservoir host first identified in 1993.

■ TRANSMISSION
By inhalation of dried rodent urine or saliva.

■ PORTAL OF ENTRY
Upper respiratory tract.

■ INCUBATION PERIOD
3–10 days.

This is a recently emerged disease. The first case occurred in a Navajo Indian who had been visiting caves in the Four Corners region (where four states in the southwest U.S.A. join) in 1993. The patient died from very severe respiratory tract infection. By August 1995 there had been a total of 115 cases with a 51 percent mortality rate.

By fall 1995, a brand new hantavirus, sin nombre virus (*sin nombre* means "no name" in Spanish) had been confirmed as the cause, and the reservoir host as the deer mouse. The disease caused by the sin nombre virus is called hantavirus pulmonary syndrome (HPS), and has emerged throughout the Americas.

CLINICAL FEATURES

Patients present with fever, myalgia, a cough, and difficulty breathing. The lungs then fill with fluid (pulmonary edema), and progression to death is rapid (2 to 16 days after onset).

TREATMENT

Patients need intensive-care support, and rib-

RNA VIRUSES: RHABDOVIRUS
rabies virus/rabies

Although rabies virus is rare in Europe and North America, it is still a serious problem, in terms of both human health and economic progress, in large parts of Africa, Asia, and South America. It is zoonotic, and estimated to cause 35,000 deaths each year, largely through dog bites.

CLINICAL FEATURES
Patients develop malaise, headache, anorexia (aversion to food), nausea, fatigue, and sometimes fever, before entering the acute neurologic phase, when diagnosis is usually made. The "furious" form is characterized by hyperactivity, hallucinations, and bizarre behavior.

Hydrophobia, popularly thought to be the most distinctive feature of rabies, is found in fewer than 50 percent of cases. It consists of painful spasms in the throat while swallowing, causing a fierce aversion to taking water or food.

The "dumb", or "paralytic," form can occur by itself or follow the furious form. Both forms result in coma, and all who enter the neurologic phase die.

TREATMENT
No specific treatment, but postexposure prophylaxis by giving rabies vaccine and pooled antirabies immunoglobulin is beneficial.

PREVENTION
The first vaccine was developed by Louis Pasteur in 1885. Current vaccines are much safer and more effective. Because the incubation period of rabies can be prolonged, it is one of the few infections for which immunization can be given after the virus has been acquired.

QUICK REFERENCE

■ **ORIGIN**
An animal virus accidentally transmitted to humans.

■ **FIRST IDENTIFIED**
The first probable description of rabies was in the Eshunni code, and dates from 2200 B.C. The virus was cultured in 1948.

■ **TRANSMISSION**
Via the saliva of infected animals, usually through bites.

■ **PORTAL OF ENTRY**
Through the skin by biting, or inoculation into cuts and grazes.

■ **INCUBATION PERIOD**
As short as 9 days or as long as several years.

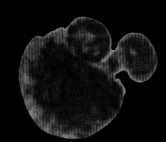

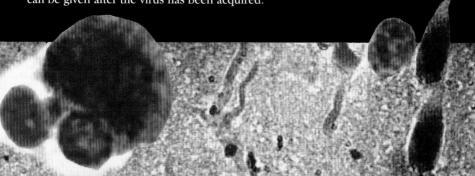

RNA VIRUSES: FILOVIRUS

ebola/marburg/
viral hemorrhagic fever

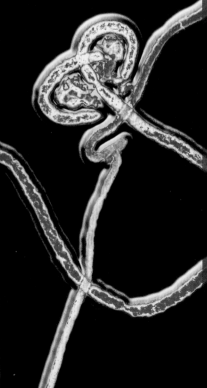
These viruses are recently emerged members of the filovirus family. They are RNA viruses which can grow up to half an inch (14 mm) in length and appear to the naked eye as long fine threads (*filo* is Latin for thread).

FIRST DESCRIPTION

The first known human filovirus infections were described in Marburg in Germany in 1967 and shortly afterwards in Frankfurt, Germany, and Belgrade, Serbia. The common factor was direct or indirect contact with blood or tissue from vervet monkeys that had been imported from Uganda via London. In all, there were 32 cases of infection, which presented as viral hemorrhagic fever (VHF), and killed seven people.

EARLY AFRICAN CASES

In 1975, a further three cases emerged in South Africa, one of which resulted in death. Then, in 1976, two large outbreaks of VHF occurred, the first in Nzara, Southern Sudan, and the other in Yambuku, Democratic Republic of Congo (formerly Zaire), near to the Ebola River. In total, there were 284 cases and 148 deaths in Sudan, and 318 cases and 280 deaths in Congo. The viruses responsible were called Ebola but it soon became clear that there were slight differences between some of them, so they are now called Ebola/Zaire and Ebola/Sudan.

HUMAN INFECTION

In 1989, in the U.S. town of Reston, Virginia, monkeys imported from the Philippines developed fatal hemorrhagic fever and a filovirus was isolated. There was much concern that humans in contact with the monkeys might develop VHF. However, although humans became infected they did not develop the disease. This virus was called Ebola/Reston. A number of outbreaks of VHF have occurred in the U.S.A., Italy, Côte d'Ivoire, Gabon, Congo, and

Uganda, and in 1994, a new strain, Ebola/Côte d'Ivoire, caused VHF in a patient who had conducted an autopsy on a dead chimpanzee.

INTERNATIONAL SPREAD
A doctor looking after patients in Gabon became infected, and while incubating the disease, flew to Johannesburg. Having been admitted to hospital there, he infected a nurse, who died. This illustrates the concern that modern air travel can spread the virus. In 2000, the largest outbreak to date occurred in Gulu, Uganda, where 400 patients developed disease and 160 died.

CAUSES OF INFECTION
Although monkeys have been implicated, they are unlikely to be the reservoir host since VHF kills them, too. Ebola can also infect bats and, possibly, small rodents (though asymptomatically). Doctors and nurses looking after patients with VHF are also at risk, as the virus is transmitted from person to person via blood-stained body fluids and tissues. However, in the original Marburg outbreak, one patient infected his wife via sexual intercourse 76 days after he had become ill.

CLINICAL FEATURES
Rapid onset of fever, malaise, muscle pains, and severe headaches, followed by nausea, vomiting, and bloody diarrhea. Within five to seven days the patient shows evidence of bleeding into skin and mucous membranes. Death usually occurs after six to nine days. Mortality rates for Ebola are between 50 percent and 80 percent, whereas for Marburg, rates are between 33 percent and 57 percent.

TREATMENT
Supportive therapy is all that can be offered.

PREVENTION
No vaccine is available.

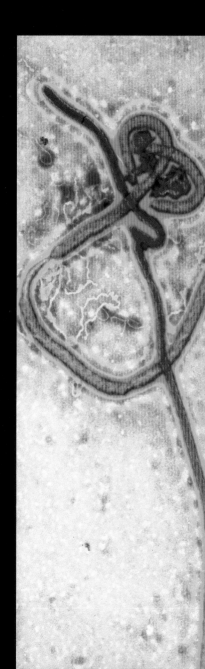

RNA VIRUSES: LYSSAVIRUS

bat lyssavirus/rabies

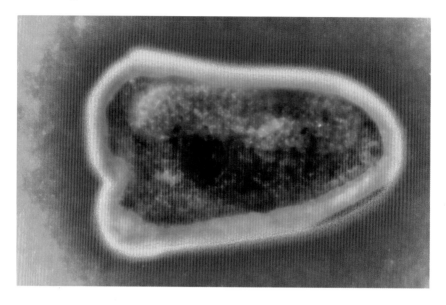

■ **ORIGIN**
Naturally occurring bat viruses.

■ **FIRST IDENTIFIED**
In the U.S.A. in 1994.

■ **TRANSMISSION**
From bat bites.

■ **PORTAL OF ENTRY**
Through the skin, to the brain via peripheral nerves.

■ **INCUBATION PERIOD**
20–60 days.

These viruses are distinct from rabies virus and, so far, at least six different genotypes have been found (genotype 1 is a rabies virus). Those in Africa tend to be genotypes 2 and 4, in Europe, genotypes 5 and 6, and in the U.S.A., genotype 1.

Research in the U.S.A. into how patients who had no contact with the usual animal hosts of rabies virus developed the disease revealed that carnivorous bats also transmit the classical rabies virus. In addition, the remaining bat lyssaviruses are carried and transmitted by both insectivorous and fruit bats, though these lyssavirus infections are rare.

CLINICAL FEATURES
The features are as described for classical rabies (see page 53).

TREATMENT
No specific antiviral therapy is available. Post-exposure prophylaxis, as with rabies, is possible.

PREVENTION
Current rabies vaccines should be protective.

RNA VIRUSES: RETROVIRUSES
human T-cell leukemia virus 1/
tropical spastic paresis and leukemia

Human T-cell lymphotropic virus 1 (HTLV-1) is a retrovirus, that is, one that inserts a DNA copy of its genome into the host cell in order to replicate. The virus persists for the lifetime of the infected individual. Retroviruses pathogenic to humans include HTLV-1 and HIV. HTLV-1 is associated with two disease manifestations: adult T-cell leukemia and tropical spastic paresis, but most infected individuals have no disease.

The virus is highly endemic in Japan and the Caribbean. Natural transmission appears to be from mother to baby via breast-feeding, but it can also be transmitted by needle sharing, blood transfusion, and organ transplantation.

CLINICAL FEATURES
Adult T-cell leukemia occurs in between 2 percent and 4 percent of those infected with HTLV-1 in endemic areas, usually after a latent period of several decades. Tropical spastic paresis is characterized by chronic degeneration of motor nerves from the brain down the spinal cord, leading to rigid or spastic paralysis of muscles. Fewer than 1 percent of HTLV-1-infected individuals develop this disease.

TREATMENT
It is not clear whether or not the antiretrovirals used to treat AIDS are beneficial.

PREVENTION
No vaccine is available.

QUICK REFERENCE

■ **ORIGIN**
Unknown.

■ **FIRST IDENTIFIED**
Virus first isolated in 1980.

■ **TRANSMISSION**
From mother to baby via breast-feeding, or blood-to-blood through needle sharing, transfusion, or organ transplantation.

■ **PORTAL OF ENTRY**
Mouth, or through the skin by injection.

■ **INCUBATION PERIOD**
Decades.

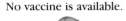
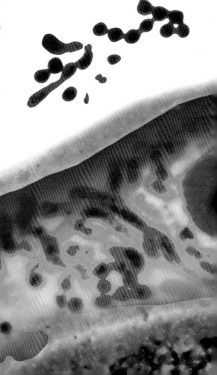

RNA VIRUSES: RETROVIRUSES

human immunodeficiency viruses
1 & 2/AIDS

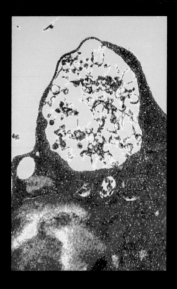

In 1981 a "new disease" was reported from the U.S.A. It consisted of unusual respiratory tract infections occurring in young, previously healthy, gay men. It was termed acquired immunodeficiency syndrome, or AIDS; *acquired* because it was not congenital, *immunodeficiency* because it was presenting with diseases that only occur when the immune system is malfunctioning, and *syndrome* because it was a collection of all these features.

The agent causing it was found in 1983 and was called human immunodeficiency virus 1 (HIV-1). HIV-2 is a related virus that was initially found in West Africa in 1986.

TRANSMISSION
HIV mutates very rapidly (at a rate of one in 1,000) and so is constantly altering its antigenic nature to avoid immune recognition and elimination. Once infected, a person is infected for life, and if untreated, the rate of progression to AIDS in gay men is 50 percent in five years.

HIV is acquired at birth or by breast-feeding from mother to baby, by hetero- or male homosexual activity, and by blood-to-blood transmission such as accidental needle-stick in health workers, needle-sharing, transfusion of blood or blood products, and organ transplantation.

AREAS OF ATTACK
Once in the blood, HIV attacks, in particular, T-helper cells. These cells are involved in optimal antibody production and contribute to cell-mediated immunity, which are the two main arms of the immune system. During the initial phase of infection, approximately 100 million new virus particles are produced each day and a million T-helper cells are destroyed. Eventually, virus production and destruction of T-cells slows. Patients can stay in this latent phase showing no disease for many years. However, the virus reactivates rapidly and the T-cells are destroyed, leading to infections.

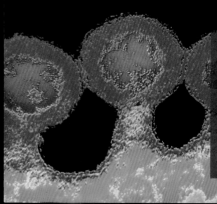

STATISTICS

It is estimated that there are 42 million people worldwide infected with HIV, and this number is expected to more than double by 2010; 61 percent of cases are in sub-Saharan Africa.

CLINICAL FEATURES

Following initial infection, around 5 percent of individuals develop glandular-fever-like illness. Patients remain asymptomatic for years after this. Then, prior to full-blown AIDS, some develop enlarged lymph glands. Once they have AIDS, patients develop infections to which they would normally be immune, and which prove fatal.

TREATMENT

There are now many drugs that inhibit the growth of HIV and reconstitute the immune system. However, as HIV is highly mutable, viral resistance is a serious ongoing problem.

PREVENTION

No vaccine is currently available. Public health education, such as promoting the use of condoms, has helped decrease transmission, and mother-to-baby transmission can also be decreased by giving the mother antiretrovirals.

QUICK REFERENCE

■ **ORIGIN**
HIV-1 is derived from a chimpanzee virus, and HIV-2 from a macaque virus.

■ **FIRST IDENTIFIED**
The first publication of AIDS was in 1981. HIV-1 was isolated in 1983 and HIV-2 in 1986.

■ **TRANSMISSION**
Mainly via sexual intercourse but a number of cases are transmitted from mother to baby at birth or by breast-feeding. Blood-to-blood transmission is now less common, except when caused by needle sharing.

■ **PORTAL OF ENTRY**
The bloodstream via skin or mucous membranes, or directly by injection.

■ **INCUBATION PERIOD**
Several years.

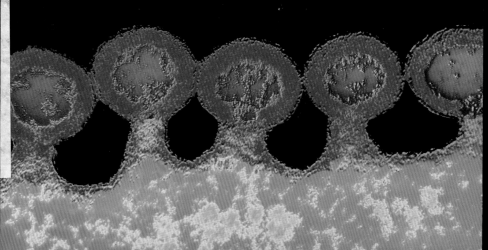

RNA VIRUSES: CORONAVIRUSES

human coronavirus/respiratory tract infection, diarrheal disease

QUICK REFERENCE

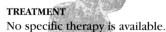

■ ORIGIN
Unknown.

■ FIRST IDENTIFIED
Viruses were first cultured
in the early 1960s, and the family
given its name in 1968.

■ TRANSMISSION
From respiratory spray via
coughing and sneezing.

■ PORTAL OF ENTRY
Nasal epithelium.

■ INCUBATION PERIOD
3 days.

On their surface, coronaviruses have widely spaced club-like projections, which give the family its name—the characteristic appearance supposedly resembles the solar corona (*corona* is Latin for crown) seen during an eclipse.

There is a large number of coronaviruses, most of which infect animals other than humans. Prior to 2003, there were three known genogroups of coronaviruses: two of these contained mammalian (including human) coronaviruses, whereas the third contained only avian (bird) viruses.

Human coronaviruses are a major cause of upper respiratory tract infection. It is assumed they are spread by respiratory spray generated by coughing or sneezing. The virus grows best at around 92°F (33°C), which is the temperature of the nasal passages. Infection appears to be limited to the nasal epithelium.

Infections peak in winter and early spring in temperate countries. It is estimated that between 10 percent and 20 percent of the population become infected each year, though in some years, this can rise to 35 percent of children and adults.

CLINICAL FEATURES
Patients have a typical "cold," with a runny nose, nasal congestion, fever, and a sore throat. The illness usually lasts no longer than seven days. Coronaviruses have also been demonstrated to be a rare cause of diarrheal disease.

TREATMENT
No specific therapy is available.

PREVENTION
No vaccine is available.

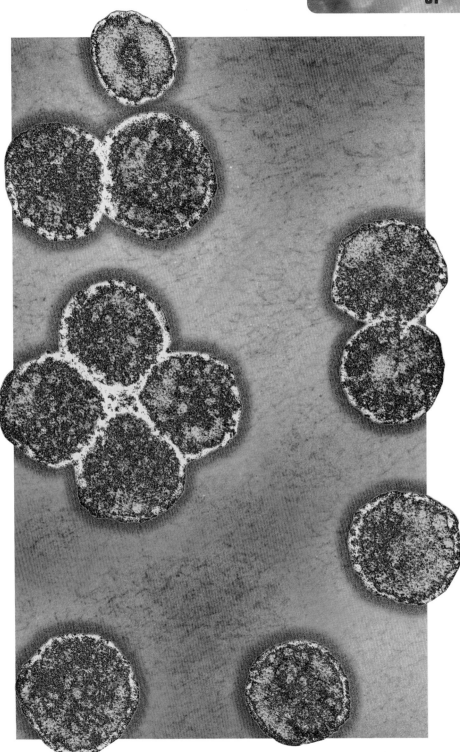

RNA VIRUSES: CORONAVIRUSES

SARS coronavirus/
severe acute respiratory syndrome

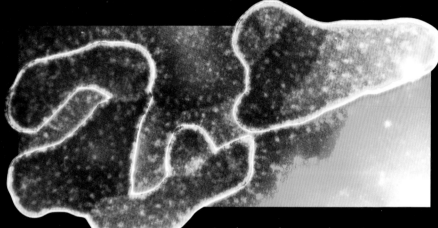

QUICK REFERENCE

■ **ORIGIN**
A coronavirus of civet cats accidentally transmitted to humans.

■ **FIRST IDENTIFIED**
In China in 2002. The virus was isolated in 2003.

■ **TRANSMISSION**
Person-to-person via respiratory secretions or on hands.

■ **PORTAL OF ENTRY**
Upper respiratory tract.

■ **INCUBATION PERIOD**
5–8 days.

In October and November 2002, reports emerged of a highly contagious, very severe lower respiratory tract infection in Guandong Province in southern China. The reports were largely overlooked until March 2003 when patients from the area, or people who had traveled from Guandong, presented symptoms of the infection in Hong Kong, Singapore, and Canada. The World Health Organization issued a global alert, and by May 2003 there had been a total of 8,240 probable cases of the disease, with 745 deaths, affecting a total of 31 countries worldwide.

THE ARRIVAL OF SARS
The disease was named severe acute respiratory syndrome, or SARS. Initially the major problem was in distinguishing SARS from other infections causing similar diseases. For example, influenza, respiratory syncytial virus, human metapneumovirus, and *Legionella pneumophila* all produce similar symptoms. However, in early April 2003, groups from Hong Kong, Germany, and the U.S.A. reported the isolation of a new virus, now called SARS-coronavirus, and it has now been proven that this new virus is the cause of SARS.

recent jump from its animal host, the civet cat, which along with other wild animals, is sold as food in China's markets. To date, up to 40 percent of market workers have been shown to have been infected with SARS.

CLINICAL FEATURES

Patients present with a fever, chills, malaise, aversion to food (anorexia), myalgia (muscle pain), headache, and a cough. On average they require hospital treatment three to five days after onset. With the correct treatment, the fever drops after a further three to five days but in those with very severe disease, it returns at day nine, along with acute watery diarrhea and a worsened chest infection (20 percent of patients requiring artificial ventilation).

The mortality rate in those over 60 is 43 percent, and in those under 60, 13 percent. For those who survive, the average duration of hospitalization is three weeks. The illness in children has been much milder, rarely requiring hospitalization.

TREATMENT

No specific antiviral drugs are available as yet, but combinations of interferon-alpha, ribavirin, and steroids may be beneficial.

PREVENTION

No vaccine is available.

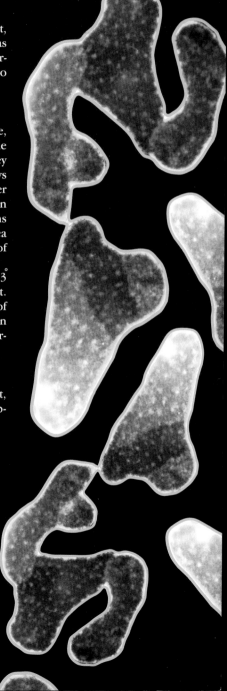

DNA VIRUSES: PARVOVIRUS

human parvovirus/
erythema infectiosum

QUICK REFERENCE

- **ORIGIN**
Unknown.

- **FIRST IDENTIFIED**
Virus first described in 1975 and
associated with disease in the 1980s.

- **TRANSMISSION**
Respiratory secretion.

- **PORTAL OF ENTRY**
Upper respiratory tract.

- **INCUBATION PERIOD**
6–11 days.

Parvovirus is the smallest of the DNA viruses. Only one member of the parvovirus family, the genus erythrovirus (sometimes called parvovirus B19), can infect humans. Infection occurs throughout the year and has a worldwide distribution. However, epidemics do occur in primary schools, usually in the spring when up to 40 percent of children may develop disease.

The virus is transmitted via respiratory secretions but blood-borne transmission can also occur as the level of viremia (virus in the blood) is high and can persist for some time. In particular, this virus targets the precursor cells that make our red blood cells. It can cross the placenta and cause fetal damage if the mother acquires infection for the first time during pregnancy.

CLINICAL FEATURES

In children and adult males, the infection is usually a short febrile illness. However, in some children it also produces a maculopapular (flat and raised) erythematous (red) rash on the body and a characteristic "slapped-cheek" appearance.

In adults, and particularly females, a maculopapular rash can also be accompanied by joint pains and swelling. If the infected patient has hemolytic anemia (low levels of red cells in the blood due to their destruction), this precipitates a rapid worsening of anemia (aplastic crisis), which is life-threatening.

When a fetus is infected in the womb, in a small proportion of cases it develops heart failure because it has become anemic. This results in fluid overload in the tissues and the fetus is at risk of dying.

TREATMENT

No antiviral drugs are available.

PREVENTION

No vaccine is available.

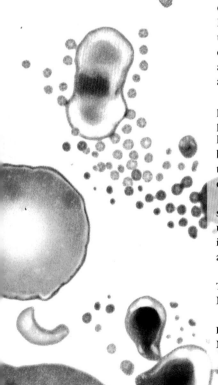

DNA VIRUSES: PAPOVAVIRUSES

polyomavirus/progressive multifocal leukoencephalopathy

The papovirus family consists of two genera: the polyomaviruses and the papillomaviruses. Two of the polyomaviruses infect humans, and these are called JC and BK after the initials of the patients from whom they were isolated. Most individuals worldwide are infected with one or both viruses. Infections occur in childhood and the virus seems to remain following initial infection, either latently or persistently.

During periods of immunosuppression, or even pregnancy, virus excretion restarts, usually into urine. It is not certain whether or not BK virus causes any disease, but it has been found in a minority of childhood respiratory tract infections. JC virus can cause progressive multifocal leukoencephalopathy (PMFL), but only in immune-suppressed individuals.

CLINICAL FEATURES

PMFL is a very rare disease that was first described in 1958. It results from long-lasting immune suppression, such as malignancy (Hodgkin's lymphoma, lymphatic leukemia, and AIDS). Patients suffer increasing loss of mental function and neurological control. Progression to death takes between three and eight months.

TREATMENT

No specific antiviral treatment is available.

PREVENTION

No vaccine is available.

QUICK REFERENCE

■ **ORIGIN**
Unknown.

■ **FIRST IDENTIFIED**
Progressive multifocal leukoencephalopathy was first described in 1958. JC and BK viruses were first isolated in 1971.

■ **TRANSMISSION**
Probably via respiratory secretion.

■ **PORTALS OF ENTRY**
Ingestion or inhalation.

■ **INCUBATION PERIOD**
Unknown.

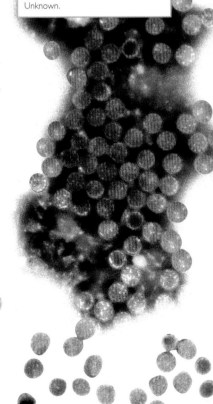

DNA VIRUSES: PAPOVAVIRUSES

human papillomavirus/
warts, carcinoma of the cervix

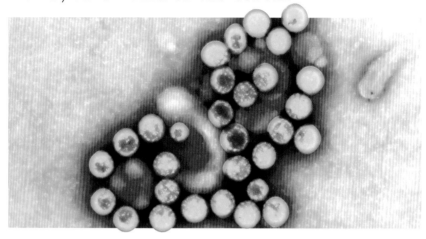

QUICK REFERENCE

■ **ORIGIN**

Virus appears to have evolved with humans.

■ **FIRST IDENTIFIED**

Warts have been described from ancient times, and papillomaviruses were first described in 1933. HPV was not associated with carcinoma of the cervix until 1983.

■ **TRANSMISSION**

Direct or indirect contact with infected skin or mucosa.

■ **PORTALS OF ENTRY**

The skin or mucous membranes.

■ **INCUBATION PERIOD**

Skin warts: up to 2 years. Genital warts: around 6 weeks. Carcinoma of the cervix: 4–20 years.

There are more than 100 different genotypes of human papillomaviruses (HPV). Some cause common warts, others genital warts, plantar warts, and cervical cancer. All are spread by direct or indirect contact with shed skin. Genital warts and cervical carcinoma are spread by sexual intercourse.

CLINICAL FEATURES

Cutaneous warts on the soles of the feet become buried in the skin and are called verrucas. Warts elsewhere occur as discrete single or multiple raised craggy lumps. Eventually, when the immune system recognizes the virus-infected cells, a ring of inflammation develops around the base of the wart and it falls off. Genital warts have a high attack rate. Carcinoma of the cervix follows some years after infection.

TREATMENT

Surgical removal, freezing, cauterization, or podophyllotoxin treatment. None achieves a complete cure. Cervical cancer requires surgical removal with irradiation or chemotherapy.

PREVENTION

No vaccine, but transmission is greatly reduced by covering warts with dressings or a condom.

DNA VIRUSES: ADENOVIRUS

adenoviruses/respiratory tract infection, rashes, diarrheal disease, conjunctivitis

To date, over 51 different human adenoviruses have been described. The different viruses are able to grow in the human respiratory tract, gastrointestinal tract, and conjunctivae of the eye. Less commonly, they can grow in the urinary bladder and liver, and very rarely cause infection of the pancreas, heart, or central nervous system. However, many infections are subclinical and in some cases can persist for years.

Adenovirus 40/41 are a minor cause of gastroenteritis and are spread feco-orally. Adenovirus 7 can cause severe pneumonia and is spread via respiratory secretion. Adenovirus 5 causes epidemic conjunctivitis and is probably spread on hands and via respiratory secretion. Each of the adenoviruses is found worldwide, and some are able to remain latent.

CLINICAL FEATURES

Approximately 5 percent of childhood acute respiratory diseases are caused by adenovirus. Adenoviruses 4 and 7 cause outbreaks of respiratory tract infection (RTI) in military recruits, and adenovirus 5 can also cause a whooping-cough-like illness in children.

Adenoviruses 3 and 7 can cause waterborne outbreaks of conjunctivitis. Epidemic keratoconjunctivitis involves both the cornea and conjunctivae and is much more severe. It has been called "shipyard eye" because it spread among shipyard workers who shared goggles.

Adenoviruses 40/41 cause a mild diarrheal disease, usually in children.

TREATMENT

No antiviral therapy is available.

PREVENTION

Oral live attenuated adenoviruses 4 and 7 vaccines are used in the U.S. military.

QUICK REFERENCE

- **ORIGIN**
  Probably evolved with humans.

- **FIRST IDENTIFIED**
  First human adenovirus grown in 1953.

- **TRANSMISSION**
  Via respiratory secretion, contact, or feco-oral spread depending upon the adenovirus and site of infection.

- **PORTALS OF ENTRY**
  The eyes, mouth, or upper respiratory tract.

- **INCUBATION PERIOD**
  2–7 days.

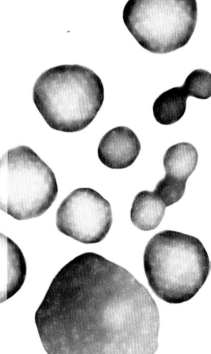

DNA VIRUSES: HERPESVIRUSES

herpes simplex virus type 1/
cold sores

■ **ORIGIN**

Has evolved with humans over 8–10 million years.

■ **FIRST IDENTIFIED**

The term herpes has been used in medical literature for centuries and is derived from the Greek word meaning to creep. The disease was linked to a viral cause in 1921.

■ **TRANSMISSION**

Through saliva or direct contact with cold sores.

■ **PORTALS OF ENTRY**

The skin and mucous membranes.

■ **INCUBATION PERIOD**

Not applicable.

Herpesviruses form a large family of over 130 viruses. Humans are natural hosts for eight of these and can be infected accidentally, and fatally, by one more: herpesvirus simiae (or simian B virus). A key characteristic of the human herpesviruses is that once an individual has been infected, he or she is infected for life.

Herpes simplex virus type 1 (also known as human herpesvirus 1, or HHV-1) is usually acquired in early childhood and can remain latent for long periods. However, the virus can reactivate sporadically, resulting in a cold sore.

CLINICAL FEATURES

Feelings of numbness or tingling around the mouth often indicate the beginnings of a cold sore. A raised red lump then appears, and this rapidly turns into a blister, which eventually bursts and crusts over before the scab falls off. The whole process takes about a week.

The virus can also enter through the eyes, causing conjunctivitis. Recurrent infections of the cornea cause scarring and this is the most common infective cause of blindness in developed countries. Rarely, the virus causes encephalitis.

TREATMENT

A number of safe and effective antiviral drugs are available, including Aciclovir, an ointment applied to cold sores that decreases the length of illness by about two to three days.

PREVENTION

No licensed vaccine is available.

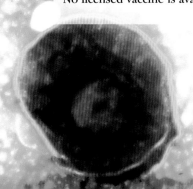

DNA VIRUSES: HERPESVIRUSES

herpes simplex type 2/
genital herpes

In general, herpes simplex type 2 (also known as human herpesvirus-2, or HSV-2) causes infections below the waist. Most often it is a sexually transmitted infection, termed genital herpes. The virus is inoculated into skin or mucous membrane of the genital tract. Though it remains latent for long periods, the virus reactivates regularly, causing blisters to form on the genital area. Once infected, a patient is infected for life. However, not all of those with the latent virus experience recurrences.

There has been an inexorable rise in incidences of genital herpes infections over the last three decades. The lesions of genital herpes are uncomfortable, can be painful, and are unsightly, though not life-threatening.

However, if a baby is born when genital herpes lesions are present in the mother's birth canal, the newborn is at risk of severe neonatal herpes. Approximately one in 1,000 pregnancies in the U.S.A. are complicated by this. In almost half the cases, the virus spreads from the baby's skin to the brain (encephalitis), or to other organs including the liver and lungs. The latter infection, if untreated, has a 70 percent mortality rate.

CLINICAL FEATURES
The primary infection can be severe and, if so, frequent recurrence is much more likely. Both primary and secondary lesions present single or multiple vesicles in the genital region. Primary lesions are often also accompanied by fever, muscle pains, and malaise. This is less likely with secondary lesions. Recurrences can be as many as 20 per year.

TREATMENT
Oral Aciclovir is useful.

PREVENTION
No vaccine is available.

QUICK REFERENCE

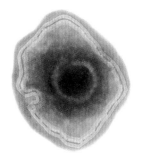

■ **ORIGIN**
Has evolved with humans over the last 8–10 million years.

■ **FIRST IDENTIFIED**
Virus first properly characterized in the 1960s.

■ **TRANSMISSION**
By close physical or sexual contact.

■ **PORTALS OF ENTRY**
Through the skin or mucosa of the genital tract.

■ **INCUBATION PERIOD**
Primary infection: 2–12 days.

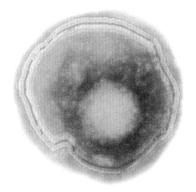

DNA VIRUSES: HERPESVIRUSES

varicella zoster virus/
chickenpox

QUICK REFERENCE

■ **ORIGIN**
The first description was in the 16th century. Virus first grown in artificial culture in 1958.

■ **TRANSMISSION**
By airborne respiratory secretions or ingestion of vesicle fluid.

■ **PORTALS OF ENTRY**
Upper airways.

■ **INCUBATION PERIOD**
13–17 days.

Varicella zoster virus (VZV) is the cause of the common childhood infectious disease, chickenpox. The recurrent or secondary disease is shingles, or zoster. The chickenpox rash, which tends to be heaviest on the trunk and abdomen, appears about 21 days after infection. Patients are most infective three to five days before this.

Afterwards, the virus remains latent for several decades, but as the host's immune system ages and becomes less effective, it can reactivate, forming a belt of painful vesicles around the chest, known as shingles. Those not previously infected with VZV can catch chickenpox from patients with chickenpox or shingles. However, shingles is not transmitted person-to-person.

CLINICAL FEATURES
Chickenpox presents with red raised papules that enlarge and develop into blisters. At any one time, papules, blisters, burst blisters, and scabs will be present. The virus occasionally causes encephalitis and even pneumonia. Shingles usually begins with pain along the nerve in which the reactivation is occurring, followed by a flood of red spots along the distribution of the nerve. This lasts for up to 16 days, and can recur.

TREATMENT
Aciclovir is of dubious benefit in mild chickenpox but is used in severe disease. It is also beneficial in shingles if given early enough.

PREVENTION
The live-attenuated Oka vaccine is used in the U.S.A.

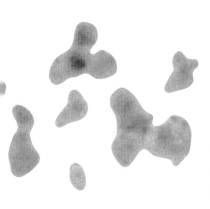

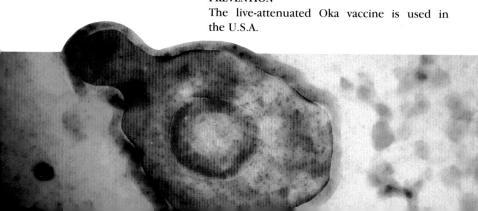

DNA VIRUSES: HERPESVIRUSES

epstein barr virus/
glandular fever, lymphomas

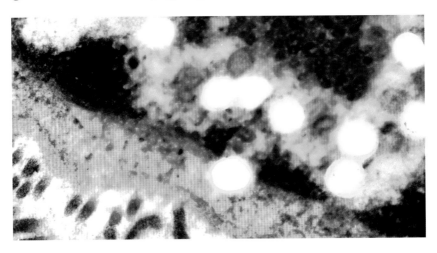

In developing countries, most children are infected with Epstein Barr virus (EBV) by the age of 10, and this is usually asymptomatic. In developed countries, infection usually occurs later and can be symptomatic. The virus causes glandular fever, or infectious mononucleosis (IM). By adulthood, up to 60 percent of us will have been infected, for life.

EBV is most often spread by exchange of saliva, which is why it is often called "kissing disease." The virus remains latent in B-lymphocytes (the cells that make antibodies) and occasionally causes them to become cancerous.

CLINICAL FEATURES
The illness begins with malaise followed by fever, pharyngitis, and sometimes frontal headaches. Lymph glands, tonsils, and spleen all become enlarged, and the tonsils will have pus on their surface. Patients may develop a punctate red rash. Occasionally, the disease can persist for over a year as chronic viral fatigue syndrome.

TREATMENT
No specific antiviral treatment is available.

PREVENTION
No vaccine is available.

QUICK REFERENCE

■ **ORIGIN**
Evolved with humans.

■ **FIRST IDENTIFIED**
Disease (IM) first described in 1929. EBV first cultured artificially and its role in IM proven in 1964.

■ **TRANSMISSION**
Mainly by exchange of saliva, but can also be transmitted via blood transfusion.

■ **PORTAL OF ENTRY**
Mouth.

■ **INCUBATION PERIOD**
30–50 days.

DNA VIRUSES: HERPESVIRUSES

cytomegalovirus/
congenital infection

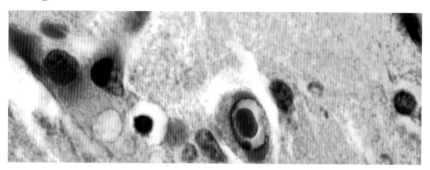

QUICK REFERENCE

■ **ORIGIN**
An ancient virus that has evolved with humans.

■ **FIRST IDENTIFIED**
Virus first isolated in 1956. The first description of cytomegalic inclusion disease was in 1881.

■ **TRANSMISSION**
By exchange of saliva, through breast milk, semen, or cervical fluid. Can also be transmitted via blood transfusion.

■ **PORTALS OF ENTRY**
Mouth or genital tract.

■ **INCUBATION PERIOD**
Infectious mononucleosis: 5–6 weeks.

In most cases, cytomegalovirus (CMV) infection occurs asymptomatically, but occasionally causes a glandular-fever-like illness, especially if acquired via blood transfusion. The virus can be transferred across the placenta, by breast milk, sexual intercourse, or saliva exchange. After childhood, the estimated infection rate is 1 percent of the population per annum. By adulthood, between 40 percent and 60 percent will have been infected. Once infected, the individual is infected for life. CMV remains as both a latent and a persistent infection, and virus reactivation and reinfection with different CMV strains is possible.

Infection can be dangerous if a woman acquires her first episode of CMV infection during pregnancy. The virus can cross the placenta to infect the growing fetus, causing mental retardation. When CMV infects immune-compromised patients, such as those receiving bone marrow or renal transplants, or those with AIDS, it can cause hepatitis, enteritis, pneumonia, or encephalitis.

CLINICAL FEATURES
Most infections are asymptomatic.

TREATMENT
Ganciclovir and Foscarnet are used to treat severe CMV infection in patients whose immune systems are compromised.

PREVENTION
No vaccine is available.

DNA VIRUSES: HERPESVIRUSES

human herpesvirus 6/
exanthem subitum

HHV-6 has only recently been described and, as such, it does not have a common name. It was first isolated from B-lymphocytes from a patient with AIDS in 1986. However, it has no role in the progression of HIV to AIDS. It has subsequently been shown to be a common infection in childhood and approximately 90 percent of infants are infected by the age of two. The virus has been detected in saliva, breast milk, and genital secretions. HHV-6 remains latent and persistent in the T-lymphocytes.

There are two variants of HHV-6: A and B. HHV-6A has no clear disease associations but HHV-6B is the cause of *exanthem subitum*, which is also called *roseola infantum*. The virus frequently reactivates, usually during febrile illnesses, but whether there is a reactivation disease is as yet unclear.

CLINICAL FEATURES
Exanthem means a rash on the skin (*ex* means outside, *anthem* means a flower, and *subitum* means sudden). The illness usually occurs in infants in the first year of life. The infant experiences three to five days of swinging fever then suddenly the temperature settles and a maculopapular (mixed flat-and-raised) rash appears all over the body. The virus can also provoke febrile convulsions.

TREATMENT
No specific antiviral therapy is available.

PREVENTION
No vaccine is available.

QUICK REFERENCE

■ **ORIGIN**
Has evolved with humans.

■ **FIRST IDENTIFIED**
The disease *exanthem subitum* was first distinguished from German measles in 1910. Virus was first isolated in 1986.

■ **TRANSMISSION**
By exchange of saliva, through breast milk, or genital secretions.

■ **PORTAL OF ENTRY**
Mouth.

■ **INCUBATION PERIOD**
5–15 days.

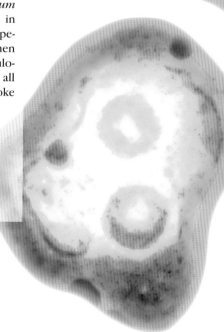

DNA VIRUSES: HERPESVIRUSES

human herpesvirus 7/
exanthem subitum

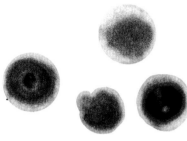

This virus, which is related to both HHV-6 and CMV, was first isolated in 1990 from the T-lymphocytes of a normal, healthy individual. HHV-7 is excreted in saliva, breast milk, and genital secretions, and infection is most often acquired through breast-feeding or kissing. Most HHV-7 infections seem to occur in those with HHV-6, usually in the second and third years of life. By adulthood, up to 80 percent of the population is infected. HHV-7 stays as a latent and persistent infection of T-helper cells (the same cells that are infected by HIV).

HHV-7 is associated with some cases of *exanthem subitum* (see page 73) but in most cases, infection is asymptomatic. As is the case with HHV-6, HHV-7 reactivates during febrile illnesses. However, no reactivation disease has yet been demonstrated.

QUICK REFERENCE

■ **ORIGIN**
Evolved with humans.

■ **FIRST IDENTIFIED**
Virus first isolated in artificial culture in 1990.

■ **TRANSMISSION**
By exchange of saliva, through breast milk, or genital secretions. It is occasionally transmitted via blood transfusion.

■ **PORTAL OF ENTRY**
Mouth.

■ **INCUBATION PERIOD**
5–15 days.

CLINICAL FEATURES
Most cases of infection are asymptomatic. In those with *exanthem subitum* the illness is exactly as that described for HHV-6B.

TREATMENT
No antiviral therapy is available.

PREVENTION
No vaccine is available.

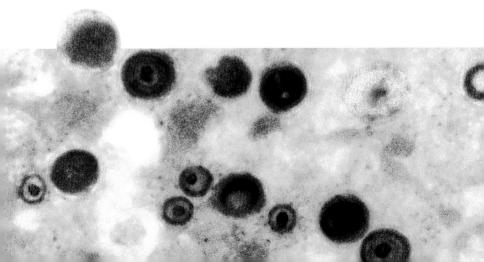

DNA VIRUSES: HERPESVIRUSES

human herpesvirus 8/
kaposi's sarcoma

The HHV-8 virus, also known as Kaposi's sarcoma-associated herpesvirus (KSHV), was discovered in 1994. Kaposi's sarcoma (KS) was first described in 1918 as a rare tumour found in individuals of Mediterranean descent. African endemic Kaposi's sarcoma is a much more invasive form, which is also quite rare.

After 1981, incidences of KS increased markedly due to the AIDS epidemic. KS is not really a sarcoma (a focal tumor of mesothelial cells) but a multifocal overgrowth of the cells lining the blood vessels and lymphatics. The growths are a characteristic chocolate brown, probably because they trap red blood cells in the spindles and the hemoglobin changes color. HHV-8 is a necessary prerequisite for the development of KS of any form, but other cofactors may be necessary. It can be found in saliva, breast milk, and genital secretions, and appears to persist in B-lymphocytes.

There is a low incidence of infection in northern Europe and North America, a higher prevalence in the Mediterranean region, and a much higher prevalence in parts of sub-Saharan Africa (over 50 percent of the population). Breastfeeding and exchange of saliva are important modes of transmission in Africa. Sexual transmission and transmission via blood transfusion are significant in Europe.

CLINICAL FEATURES
Kaposi's sarcoma develops as multiple chocolate brown or red-purple nodules. They can occur on the skin, in the lungs, digestive tract, or the brain. They are very disfiguring on the skin and can be life-threatening if in the bronchi or brain.

TREATMENT
No specific antiviral therapy. Local surgery, radiotherapy and cytotoxic drugs are used to treat the tumors.

PREVENTION
No vaccine is available.

QUICK REFERENCE

■ **ORIGIN**
Appears to have evolved with humans.

■ **FIRST IDENTIFIED**
Kaposi's sarcoma was first described in 1918. HHV-8 was first detected in 1994.

■ **TRANSMISSION**
Through saliva, breast milk, sexual intercourse, or blood transfusion.

■ **PORTAL OF ENTRY**
Mouth.

■ **INCUBATION PERIOD**
Unknown.

DNA VIRUSES: HEPADNAVIRUS
hepatitis B virus/serum hepatitis

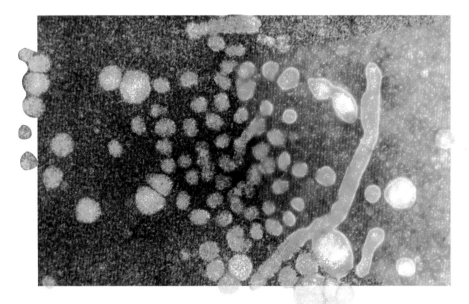

QUICK REFERENCE

■ **ORIGIN**
Probably evolved with humans.

■ **FIRST IDENTIFIED**
The earliest description of serum hepatitis was in 1885. The existence of hepatitis B virus was proposed in 1937 and the virus was first seen on electron microscopy of blood in a case of acute hepatitis B infection in 1970.

■ **TRANSMISSION**
Blood-to-blood, mother-to-baby, or by needle sharing, transfusion, or organ transplantation. Occasionally, sexually transmitted.

■ **PORTALS OF ENTRY**
Through the skin or mucosa to the liver.

■ **INCUBATION PERIOD**
2–6 months.

Hepatitis B virus (HBV) is the only member of the hepadnavirus family to infect humans. (Other hepadnaviruses infect ducks and woodchucks, but cannot infect humans.) It is solely a human pathogen and is maintained by carriers—individuals who have been infected with HBV (mostly as children) but who do not eliminate the virus.

The virus continues to grow in the carrier's liver cells and, if replication is sufficiently intense, the virus spills over into the blood. It is transmitted person-to-person via blood or through sexual intercourse.

TRANSMISSION
It is estimated that there are almost 400 million carriers worldwide, but these are not evenly distributed. In China, Southeast Asia, sub-Saharan Africa, and Greenland, up to 40 percent of the population are carriers. In the Indian subcontinent and South and Central America, between 10 percent and 15 percent of the population are carriers, and in Northern Europe, North America, and Japan the figure is less than 1 percent.

The main mode of transmission is from mother to baby at birth, but in sub-Saharan Africa, a number of transmissions occur over the first five years of life. It is also transmissible via needle sharing, sexual intercourse, tattooing, or acupuncture, and there are suggestions of insect transmission even though HBV does not grow in insects (this is mechanical rather than biological transmission). In hospitals, the risk of transfer by needle-stick accidents can be as much as 30 percent, depending on how much virus the donor has in his or her blood (figures of over 1,000 million per milliliter are possible, especially in the acute phase).

CLINICAL FEATURES

The incubation period is two to six months. In 10 percent of patients, an illness consisting of fever, joint pain, and skin rashes occurs about four weeks before the classical features of hepatitis appear. The features of hepatitis are fever, malaise, aversion to food (anorexia), jaundice, abdominal discomfort, and intolerance of fatty foods and alcohol.

Most healthy adults whose immune system is intact will only develop jaundice. However, 1 percent will develop fulminant hepatitis and die, and 10 percent will be unable to eliminate the virus and will become carriers. Of the carriers, all will eventually (after 20 to 30 years or more) develop liver failure, and between 20 percent and 30 percent will eventually develop liver cancer.

TREATMENT

Interferon and some reverse transcriptase inhibitors (for example, Lamivudine) have proved useful in some cases.

PREVENTION

A safe and effective vaccine is available.

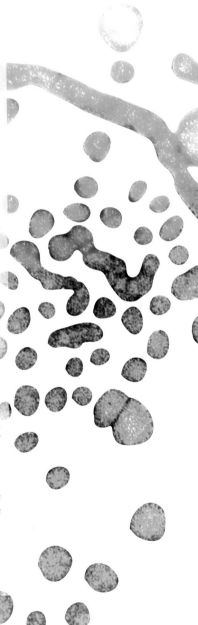

DNA VIRUSES: POXVIRUSES

smallpox virus/variola: smallpox

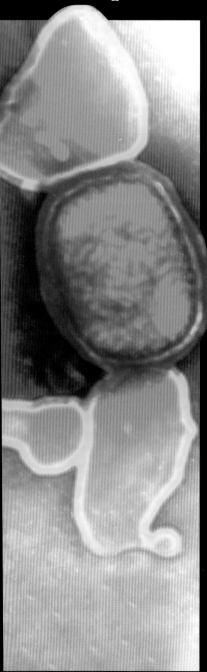

The poxviruses are a large family of DNA viruses, divided into two subfamilies: the entomopoxviruses that infect insects, and the chordopoxviruses that infect vertebrates, including humans. Of these, the most important in terms of human health is smallpox.

Smallpox is an acute infectious disease that was eliminated by a concerted World Health Organization vaccination campaign. The last naturally occurring case was in Somalia in 1977.

The origins of smallpox are unknown, but is thought to have existed in India, China, and possibly Egypt several thousand years ago. The virus came to Europe .with the Moorish invaders and was transported to the Americas by Spanish conquistadors in the 16th century. Prior to the introduction of vaccination, smallpox was responsible for between 7 percent and 12 percent of all deaths, and more than 30 percent of deaths in young children.

TRANSMISSION

The smallpox virus is solely a human pathogen and it is maintained in waves of epidemics; thus a continuous supply of those who are susceptible needs to be born. It is transmitted from person to person via respiratory secretion, although it can also be transmitted by vesicle fluid or even scabs. There is no animal reservoir of smallpox.

The virus is shed from the oropharynx of patients from the onset of rash for approximately six days. However, poxviruses are quite hardy and can survive dried on bedclothes and clothes, for example, and as scabs. Spread occurs very easily. For example, people attending the funeral of a smallpox victim in Sierra Leone spread infection to 38 other villages.

ERADICATION

The Chinese experimented in trying to prevent smallpox by inoculating material from smallpox lesions into the skin. This process, called variolation, was introduced into the U.K. by Lady Wortley Montague and the U.S.A. by

Cotton Mather in the early 1700s. However, it was as likely to produce smallpox as to prevent it. Edward Jenner's original observations of cowpox in milkmaids, and his inoculation of Edward Phipps with cowpox in 1796 to prevent smallpox, paved the way to the eventual eradication of naturally circulating smallpox.

CLINICAL FEATURES

Just prior to the rash appearing, patients experience backache, headaches, malaise, and fever. The rash then begins to appear as red papules that enlarge, vesiculate, enlarge further and become pustules, burst, and scab over. The whole process can take more than three weeks. The rash is heaviest on the arms, legs, and head. The vesicles are large and umbilicated (they have a dimple in them).

In some patients (and especially in those infected during pregnancy), there is bleeding into the vesicles (hemorrhagic smallpox). This carries a high risk of death. Overall, mortality rates range from 20 percent to over 50 percent, depending on the strain of the virus and the immunity of the population.

TREATMENT

No specific antiviral drugs were available when smallpox was circulating naturally, but drugs such as cidofavir might have been useful.

PREVENTION

The live virus vaccine (called vaccinia) derived from cowpox is highly effective. However, vaccinia is quite a "hot" vaccine with a measurable mortality rate (about six per million vaccinees), so routine mass vaccination is unlikely.

QUICK REFERENCE

■ **ORIGIN**
Probably evolved with humans but originally evolved from the cowpox virus.

■ **FIRST IDENTIFIED**
Smallpox as a disease is quite ancient and was described as far back as 3,000 years ago. The virus was first isolated in the 1930s.

■ **TRANSMISSION**
Airborne and occasionally by contact with vesicle fluid or scabs.

■ **PORTALS OF ENTRY**
Mouth and throat.

■ **INCUBATION PERIOD**
12–14 days.

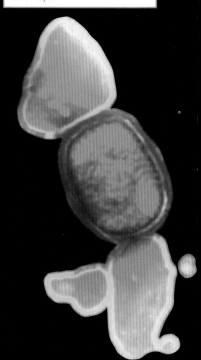

DNA VIRUSES: POXVIRUSES

monkeypox virus/monkeypox

Monkeypox is largely a tropical disease found in parts of West Africa. Although it is called monkeypox, because the disease has been found in simians who have transmitted it to humans, ground squirrels and rodents such as the giant Gambian rat, and not monkeys, are the natural host reservoirs. These hosts appear to be infected silently.

How humans become infected is not entirely clear but could be via respiratory secretion, ingestion, or by contact. For years it was thought that the spread was from animals to humans with no onward human-to-human transmission. However, in recent outbreaks in Africa and the U.S.A., person-to-person transmission has occurred.

NEW CASES
Previously found only in Africa, monkeypox spread to the U.S.A. in 2003 when someone imported giant Gambian rats as pets. The rats were housed with prairie dogs, the owner of which, after being bitten by the dogs, developed a disease with fever, malaise, chills, and a local vesicle which then spread. The virus eventually found in the lesion was shown to be monkeypox.

More than 80 people became infected, with no fatalities (although two children were admitted to hospital). Whether monkeypox has now become established in wildlife in the U.S.A. is unknown.

CLINICAL FEATURES
These are identical to those of smallpox. However, the mortality rate is lower. In African outbreaks it is around 15 percent. No one has died in the U.S. outbreak.

TREATMENT
Cidofovir might work.

PREVENTION
Vaccinia vaccination should work but has not been tested.

DNA VIRUSES: POXVIRUSES
cowpox virus/cowpox

This is the orthopox virus that was originally used by Jenner to prevent smallpox. However, molecular comparisons of cowpox and the vaccine (vaccinia virus) have now shown that the two are not the same. Cowpox is an uncommon infection limited to Western Europe, although there are descriptions of a cowpox-like illness due to cowpox-like viruses in Brazil. Cowpox is also a misnomer—cows are not the natural host reservoir of the virus.

It is now known that the reservoirs of cowpox virus are small rodents, bank voles, and woodmice, and the geographic distribution of cowpox corresponds to that of the rodents. The rodents are infected asymptomatically by cowpox for a proportion of their short lifespan, and there is some evidence that the process is actually beneficial to them.

Humans become infected by direct contact with the rodents or, more frequently, via their pet cats, which themselves have become infected from catching the rodents. Occasionally cows are infected and, for example, humans are infected by hand-milking them. Although theoretically possible, person-to-person spread is rare or even nonexistent.

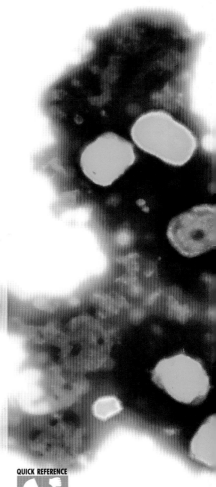

CLINICAL FEATURES
The lesions, which can be single or multiple, depending on how frequently the virus has been introduced, begin as a papule that enlarges to a vesicle, then to pustules, which then scab over. The lesions are quite florid and a large black scab can form (this can be mistaken for anthrax), which heals but leaves scars. At the onset of the lesions, patients often complain of fever, malaise, and headache.

TREATMENT
No specific therapy is available.

PREVENTION
Vaccination with vaccinia will prevent cowpox but can produce exactly the same effect (or worse) as smallpox (see pages 78–79).

QUICK REFERENCE

■ **ORIGIN**
An ancient virus of small rodents originally transmitted to humans.

■ **FIRST IDENTIFIED**
Around 1700. The virus was first grown in the 1940s.

■ **TRANSMISSION**
By contact with rodent hosts or cats.

■ **PORTAL OF ENTRY**
The skin.

■ **INCUBATION PERIOD**
5–8 days.

DNA VIRUSES: POXVIRUSES

molluscipoxvirus/
molluscum contagiosum

■ **ORIGIN**
A human virus evolved with humans.

■ **FIRST IDENTIFIED**
First published descriptions were in 1814. Jenner described it 20 years earlier but did not publish his findings.

■ **TRANSMISSION**
By contact with lesions.

■ **PORTAL OF ENTRY**
The skin.

■ **INCUBATION PERIOD**
2–7 weeks.

This virus is in a subfamily of poxviruses (molluscipox) all on its own. It has a different appearance (described as a ball of wool) and different genetic structure from the other poxviruses. Humans are its only reservoir and it is a relatively common disease of childhood, though infection can occur at any time.

It is transmitted by contact between people and between different areas of the skin on an individual. It can also be transmitted sexually. The prevalence of infection in children is between 10 percent and 20 percent with an incidence of infection of 6 percent per year. The virus has never been cultured in the laboratory.

CLINICAL FEATURES
This is a benign disease with the biggest risk being secondary infection of the lesions by bacteria. Patients present with single (unusual) or multiple raised lesions anywhere on the skin. They are usually painless and not inflamed. The lesions begin as tiny papules and enlarge to dome-shaped flaky white nodules with pearly or flesh-colored domes, which are often umbilicated. The central part, the molluscum body, can be shelled out. In most cases, the lesions resolve in six to nine months, but can take as long as four years.

TREATMENT
No antiviral drugs are available.

PREVENTION
No vaccine is available. The only preventative measure is to avoid contact with the carrier.

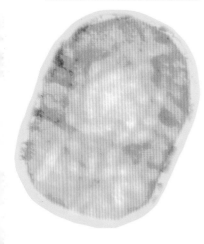

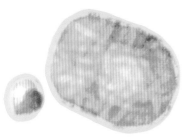

DNA VIRUSES: POXVIRUSES

parapoxvirus/
orf, milkers' nodes

When viewed through an electron microscope, parapoxviruses resemble neatly wound bobbins of string. Although two diseases originating from sheep or cattle are described, the viruses isolated from these do not differ greatly.

Orf (also called contagious pustular dermatitisis) a disease of sheep that is transmitted directly to humans. In sheep, the lesions are found in the mouth and are painful and irritating. The virus is transmitted to humans (usually farmers or vets) when they put their hands into the sheep's mouth.

In cattle, paravaccinia causes ring-shaped lesions on the teats of the udder. Humans become infected by hand-milking, which has led to the disease being called milkers' nodes. Both are classic zoonotic (transferred from animals to humans) diseases.

CLINICAL FEATURES

Lesions are most often on the fingers or hands. Both orf and milkers' nodes begin as an inflammatory papule. The orf lesions enlarge considerably (to centimeters in diameter) and are bluish granulomatous lesions. The milkers' nodes lesions tend to be smaller, highly vascularized hemispherical papules or nodules.

In each case, vesicles do not usually form but the lesions crust over and eventually heal. They are surprisingly painless. The healing process can take several weeks. Immunity is poor and repeated episodes of infection in those occupationally at risk are not uncommon.

TREATMENT

No antiviral therapy is available.

PREVENTION

No vaccine is available.

QUICK REFERENCE

■ **ORIGIN**
Viruses of animals (cattle, sheep) accidentally transmitted to humans.

■ **FIRST IDENTIFIED**
We have known about the diseases for centuries.

■ **TRANSMISSION**
From animals to humans by contact and occasionally indirectly with the virus deposited on, for example, fencing and stone walls.

■ **PORTAL OF ENTRY**
The skin.

■ **INCUBATION PERIOD**
Weeks.

2 bacteria

Bacteria are small "prokaryotes," which means that they do not possess nuclei or intracellular organelles. Unlike viruses, most bacteria are free-living and can grow in humans, other animals, and in the environment. They make up the major component of our normal bodily flora.

staphylococcus aureus/
boils, carbuncles, osteomyelitis

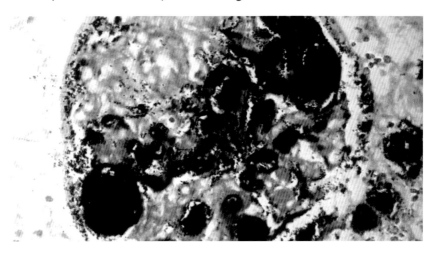

■ **ORIGIN**
An ancient bacterium.

■ **FIRST IDENTIFIED**
First described in 1880 by Sir
Alexander Ogston in Aberdeen.

■ **TRANSMISSION**
Infection can be exogenous if
transmitted from others via the
airborne route or by contact, or
endogenous when the patient's own
S. aureus causes disease.

■ **PORTALS OF ENTRY**
Through the skin or mucosa.

■ **INCUBATION PERIOD**
Varies greatly depending on disease.

Staphylococcus aureus is is part of the normal
flora of up to 40 percent of humans. It is found
in the nasal orifices, throat, and skin of the
armpits, groin, and umbilicus, where it lives
harmlessly. Infection results when it gets into
areas which are normally sterile. *S. aureus*
produces a number of toxins that include
leukocidins, which kill the white blood cells
that normally defend us, enterotoxins that
cause vomiting, and toxic shock syndrome toxins.

Diseases range from skin infections such as
boils, carbuncles, and furuncles, to pneumonia
and deep abscesses. It is also the most common
cause of osteomyelitis (infection of bone and
bone marrow) and surgical and other wounds.

CLINICAL FEATURES
Toxic shock syndrome presents suddenly with
fever, shock, and a rash. Survivors get scaling of
the skin, especially from the palms and soles.

TREATMENT
Strains of *S. aureus* have become resistant to
penicillin and other antibiotics.

PREVENTION
No vaccine is available.

staphylococcus epidermidis/
implant infection

Staphylococcus epidermidis is one of a large collection of staphylococci that differ from *S. aureus* in that they are coagulase negative, that is, not very pathogenic. They are usually harmless members of our normal flora. As its name suggests, *S. epidermidis* is particularly adapted to living on our skin. However, it can also be found in the oropharynx and intestinal tract.

Advances in medical technology have enabled these low-virulence bacteria to cause infections. They are able to adhere to materials, including metals, plastics (and other polymers), used to make implants such as venous catheters, artificial joints, heart valves, and shunts that relieve hydrocephalus (water on the brain). The *S. epidermidis* gains access to the bloodstream during implantation or at any time while the device is in place.

Once adhered to the surface of the implant, the staphylococci begin to multiply and elaborate an extracellular matrix, called slime. The bacteria, slime, and human blood proteins form what is called a biofilm. Here, the bacteria are protected from host defenses and antibiotics. Signs of infection appear only when the bacterial growth affects the function of the implant (for example, shunt blockage) or bacteria are thrown off the biofilm and block small vessels.

CLINICAL FEATURES
There are no specific features. However, if the infection is of an artificial hip joint, patients may experience fever and joint pain, and the artificial joint may become loosened.

TREATMENT
The antimicrobial susceptibility of *S. epidermidis* is not predictable, so it must be tested in vitro and therapy guided by the results. In many cases, the infected implants must be removed and replaced.

PREVENTION
No vaccine is available.

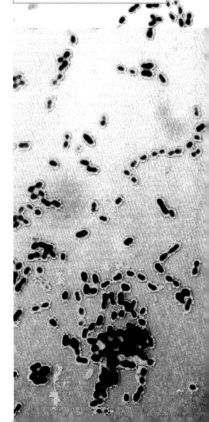

QUICK REFERENCE

■ **ORIGIN**
Unknown.

■ **FIRST IDENTIFIED**
First differentiated from *S. aureus* in the early 1900s and considered a pathogen in 1962.

■ **TRANSMISSION**
Via infection of implants.

■ **PORTAL OF ENTRY**
Through the skin to the bloodstream or cerebrospinal fluid.

■ **INCUBATION PERIOD**
Months to years.

streptococcus pyogenes/
tonsillitis, scarlet fever, skin infections

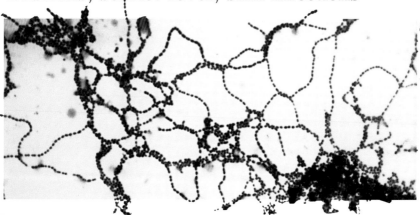

■ **ORIGIN**

Solely a human pathogen, so probably evolved with humans.

■ **FIRST IDENTIFIED**

First isolated in 1883 and split into groups in 1928.

■ **TRANSMISSION**

Person-to-person by contact, exchange of saliva, or airborne.

■ **PORTALS OF ENTRY**

The oropharynx or through abrasions and cuts in the skin.

■ **INCUBATION PERIOD**

Tonsillitis: 3–6 days.
Skin infections: 2–5 days.

The *Streptococcus pyogenes* coccus divides to produce long chains of bacteria (*streptos* is Greek for a chain). The species name *pyogenes* means pus-producing. *Str. pyogenes* is solely a pathogen of humans and is carried asymptomatically, usually in the throat, by between 20 percent and 25 percent of the population. It is spread person to person by respiratory droplets or by contact.

One of the most common infections caused by this bacterium is tonsillitis, and certain strains can release a toxin that causes the red skin in scarlet fever. Tonsillitis can occasionally (mainly in developing countries) lead to rheumatic fever, in which repeated attacks of streptococcal sore throat lead to increasing damage to the heart valves.

CLINICAL FEATURES

Skin infections include impetigo, cellulitis, and erysipelas (fever and deep-seated inflammation and pain over large areas of the skin, mainly on the limbs), and necrotizing fasciitis (an uncommon condition whereby extravirulent strains destroy the deeper tissue beneath the skin as well as the overlying skin).

TREATMENT
Penicillin.

PREVENTION
No vaccine is available.

streptococcus agalactiae/
neonatal meningitis, pneumonia, septicemia

Streptococcus agalactiae is also known as the Group B Streptococcus (GBS) and is part of the normal flora of between 20 percent and 40 percent of the population. The bacteria are carried in the gastrointestinal tract and in the vagina. Outside the neonatal period, they are rarely a cause of infection. However, in neonates (babies in the first month of life), and to a lesser extent over the next two months of life, GBS can cause severe life-threatening infection.

As the baby passes through the birth canal it picks up the mother's normal vaginal flora. If the mother is colonized by GBS, the newborn will be colonized in the mouth, intestine, and on the skin (this colonization persists for some time, and sometimes for life). In about one in 100 of these neonates, the bacteria move from their colonization sites to the bloodstream, lungs, or cerebrospinal fluid (in the brain and spinal cord).

Risk factors for development of infection include prematurity (babies born before 35 weeks of pregnancy), prolonged rupture of membranes (for longer than 24 hours), and the use of instruments such as scalp electrodes. Maternal genetic factors are also important, as some mothers cannot produce the correct antibody to GBS.

CLINICAL FEATURES
The clinical features of pneumonia and meningitis are not always present, nor is there always a fever, which makes diagnosis difficult.

TREATMENT
Penicillin.

PREVENTION
No vaccine is available. Antibiotics can be given for at-risk pregnancies.

QUICK REFERENCE

- **ORIGIN**
Unknown.

- **FIRST IDENTIFIED**
In 1883.

- **TRANSMISSION**
From mother to baby during passage through the birth canal. Can also be transmitted from mother to baby or baby to baby in neonatal intensive care units by contact, usually on hands.

- **PORTALS OF ENTRY**
Through the skin or mucosa.

- **INCUBATION PERIOD**
1–28 days.

streptococcus mutans/
dental caries, endocarditis

■ **ORIGIN**
Unknown.

■ **FIRST IDENTIFIED**
First isolated in 1924 from carious teeth, and the following year from a case of bacterial endocarditis.

■ **TRANSMISSION**
They are part of the patients' own normal flora.

■ **PORTAL OF ENTRY**
Through damage to the oral mucosa.

■ **INCUBATION PERIOD**
Indefinite.

Str. mutans is one of a large number of viridans streptococci, named so for their ability to turn the hemoglobin in red blood cells to a green color. Its main habitat is the human mouth and it is strongly linked to the development of dental caries. It is also associated with bacterial endocarditis, which occurs only where the heart valves are damaged by rheumatic fever or are congenitally malformed, where the patient has artificial heart valves inserted, or in old age when the heart valves become damaged by long-term usage.

Dental procedures such as extractions, fillings, or even scaling can result in a shower of bacteria entering the bloodstream. If the heart valves are damaged, the bacteria (frequently oral streptococci) can settle on the valve and begin to grow. Once they have grown sufficiently, the features of bacterial endocarditis become apparent. The incidence is about 20 cases per million population per year in the U.K.

CLINICAL FEATURES
Bacterial endocarditis patients experience fever, chills, and sweats. Their temperature rises and falls as showers of bacteria are released from the infected valve. Later in the disease, when clumps of bacteria are released, they may block small capillaries causing what are called splinter hemorrhages.

TREATMENT
Treatment is guided by the susceptibility pattern of the isolated bacteria.

PREVENTION
No vaccine is available, but dentists should always give prophylactic antimicrobials to patients with damaged valves.

streptococcus pneumoniae/
pneumonia, meningitis, otitis media

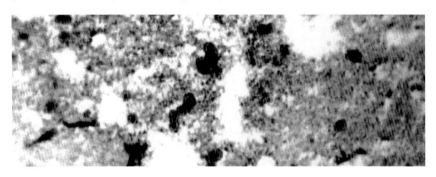

Streptococcus pneumoniae (also known as the pneumococcus) are the only major pathogens in the viridans streptococci group. They are part of the normal flora of the oropharynx, and at any time, between 20 percent and 50 percent of the population have these bacteria in their throats. It is not yet clear why pneumococci move from colonization sites to the lungs and bloodstream, converting from harmless commensals to pathogens, but viral infection may be a trigger.

CLINICAL FEATURES
Pneumococcal pneumonia can begin abruptly with a high temperature, chills, malaise, and a dry, unproductive cough. The patient will have difficulty breathing (dyspnea) and over the coming days will produce red-colored sputum that increases in volume and turns a greenish-yellow. Most deaths occur within one to three days of the illness.

TREATMENT
Though pneumococci were previously predictably sensitive to penicillin, in certain countries, including South Africa and Spain, up to 60 percent of isolates are now penicillin-resistant. Cephalosporins such as ceftriaxone are still effective in most cases.

PREVENTION
Several vaccines are available.

QUICK REFERENCE

■ **ORIGIN**
Unknown.

■ **FIRST IDENTIFIED**
First isolated by Sternberg and Pasteur, independently, in 1880.

■ **TRANSMISSION**
Part of our normal flora but can be transmitted from person to person via the exchange of saliva or coughing.

■ **PORTAL OF ENTRY**
Naso-oropharynx.

■ **INCUBATION PERIOD**
Unknown.

enterococcus faecalis/urinary tract
infection, bacteremia

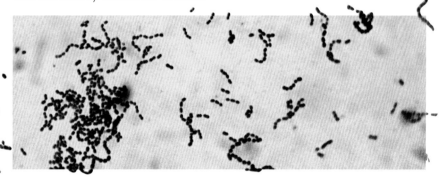

■ **ORIGIN**
Unknown, but some enterococci such as *Ent. faecium* are derived from animals.

■ **FIRST IDENTIFIED**
The first description was in France (enterocoque) in 1899.

■ **TRANSMISSION**
Most infections are endogenous, but the bacterium can be transmitted feco-orally, usually setting up colonization initially.

■ **PORTAL OF ENTRY**
The gastrointestinal tract.

■ **INCUBATION PERIOD**
Not applicable.

One of the major differences between streptococci and enterococci is that the latter are not killed by bile salts, which means they can live in the human intestinal tract. They are therefore a major component of the normal intestinal flora, where they achieve concentrations of over a million bacteria per gram of feces. They are not highly pathogenic and really only cause infections in patients whose immunity is compromised, for example, those in intensive care units.

Infection usually arises from the patient's own normal flora (though in hospital, the bacteria are spread person-to-person) and can include urinary tract infection, peritonitis, bacteremia, or endocarditis. Enterococci have a propensity to be resistant to many different antibiotics. Indeed, some recently emerged strains are resistant to almost all available antibiotics.

CLINICAL FEATURES
These vary according to the site of infection. For example, enterococci causing urinary tract infection produce fever, a burning pain on urination (dysuria) and sometimes blood in the urine.

TREATMENT
This can only be guided by how the the particular isolate reacts to antimicrobials.

PREVENTION
No vaccine is available, and it is difficult to stop it spreading in hospitals.

clostridium tetani/tetanus

Clostridium spores are very resistant to heat, drying, disinfectants, and antimicrobials and are found in large concentrations in soil, dust, and the environment generally. Clostridia are also anaerobes and *Clostridium tetani* can be part of the normal flora of the human large intestine, but this is more likely to be found in other animals such as cattle and horses.

The disease caused by *Cl. tetani* is called tetanus, and is the result of the release of a toxin called tetanospasmin, which causes muscle to go into intense and painful contractions. As little as 100 nanograms (ng) can kill an adult human.

Neonatal tetanus occurs in newborns, predominantly in developing countries, and has an 85 percent mortality rate. It is a result of contamination of the umbilical cord by instruments or dressings contaminated by *Cl. tetani* spores. As the umbilical cord stump rots, it produces an anaerobic environment that allows the spores to germinate and release the tetanospasmin.

Tetanus outside the neonatal period is worldwide. It generally occurs after a wound or scratch from an object contaminated by the bacteria.

CLINICAL FEATURES

The onset of neonatal tetanus is gradual, beginning with facial muscle spasms or spasms of the spinal muscles, causing severe arching of the back. The spasms can be precipitated by sudden noise or flashes of light. Features are the same outside the neonatal period, but may also include lockjaw. If the disease is severe, patients will be paralyzed and will require artificial ventilation.

TREATMENT

Antibiotics (penicillin) stop bacterial growth and thus the release of still more toxin. Giving antiserum to tetanospasmin is the most important intervention.

PREVENTION

The tetanus toxoid (a toxin that has been treated so that it has lost its toxic activity) vaccine is safe and highly effective.

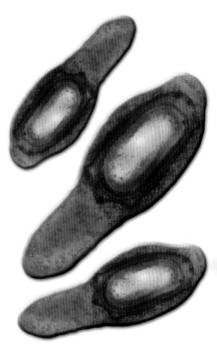

clostridium perfringens/
gas gangrene, food poisoning

■ **ORIGIN**
An ubiquitous bacterium in the environment, present as spores.

■ **FIRST IDENTIFIED**
Gas gangrene first described during World War I.

■ **TRANSMISSION**
Inoculation by instruments or through the airborne introduction of spores into necrotic wounds.

■ **PORTAL OF ENTRY**
Through the skin to wounds.

■ **INCUBATION PERIOD**
10 hours to 3 days.

Clostridium perfringens causes gas gangrene and food poisoning. Gas gangrene occurs when clostridium spores are inoculated into a deep wound where there is already tissue damage or necrosis, providing ideal conditions for the anaerobic spores to germinate (necrosis means that there is no blood supply to the area, and thus no oxygen). The spores then release toxins, which cause further tissue necrosis and release of gas.

The origin of the food poisoning form of *Cl. perfringens* disease involves contamination of meat with bacterial spores that, once ingested, eventually reach the small and large bowels and release their enterotoxin.

CLINICAL FEATURES
The disease usually occurs after accidental wounds but is occasionally associated with surgery. Onset can be within 10 hours, with sudden pain in the affected limb and swelling. The skin then becomes tense with increasing blue or black discoloration. There will be gas in the tissues that, when released, is foul smelling. If untreated, the disease progresses to toxemia and death. Patients with food poisoning experience cramping abdominal pain and profuse watery diarrhea.

TREATMENT
Penicillin plus surgical removal of the dead tissue.

PREVENTION
No vaccine is available.

clostridium botulinum/botulism

Clostridium botulinum is a spore-bearing anaerobic bacterium. It causes food poisoning but does not cause diarrheal disease. The intestine is the portal of entry for its powerful toxin (as little as 100 nanograms (ng) will kill an adult human). The bacterial spores are widely distributed in the environment, probably derived from the intestinal tract of various animal species. Botulinum toxin is a neurotoxin, which works by blocking the transmission of nerve impulses to muscle. This means that the muscles cannot contract and the patient suffers flaccid paralysis.

There are three major forms of botulism: infant botulism, wound botulism, and food botulism. Infant botulism is the rarest form and involves colonization of the infant intestine by *Cl. botulinum* following ingestion of contaminated food, such as honey. The spores germinate and release a toxin that is then absorbed from the intestine. Infection of a necrotic wound with *Cl. botulinum* spores will lead to their germination and then toxin release. Food-borne botulism occurs when food is contaminated with either spores or botulinum toxin. Foodstuffs associated with this form of the disease include vegetables, fish, fruits, and meat. Preserved food is particularly dangerous.

CLINICAL FEATURES
The first signs of botulism appear six hours after ingestion of the toxin. It begins with blurred vision, difficulty swallowing (dysphagia), and slurred speech, and is characterized by symmetrical descending (from the head downward) flaccid paralysis. Sensation remains intact.

TREATMENT
Penicillin may be of benefit in infant and wound botulism. What is most important is to give antitoxin (an antiserum). Patients may also need ventilatory support.

PREVENTION
No vaccine is available. Cooking food well enough to destroy spores will prevent food-borne disease.

QUICK REFERENCE

■ **ORIGIN**
A ubiquitous organism in the environment, present as spores.

■ **FIRST IDENTIFIED**
Botulism was recognized in outbreaks in the early 1800s.

■ **TRANSMISSION**
As a food contaminant.

■ **PORTS OF ENTRY**
Mouth or, less commonly, wounds.

■ **INCUBATION PERIOD**
6–36 hours after ingestion of toxin.

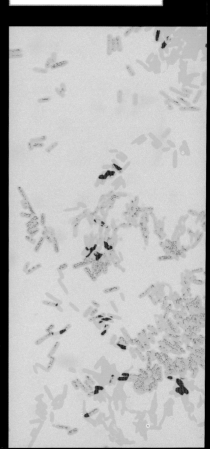

clostridium difficile/
pseudomembranous colitis

QUICK REFERENCE

■ **ORIGIN**
Bacteria derived from other humans, but spores can survive for long periods in the inanimate environment.

■ **FIRST IDENTIFIED**
Toxins and bacteria were first described in 1977.

■ **TRANSMISSION**
Can be endogenous but feco-oral transmission also occurs.

■ **PORTAL OF ENTRY**
Mouth.

■ **INCUBATION PERIOD**
Days to weeks.

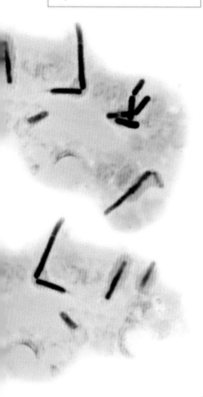

Clostridium difficile is an anaerobic, spore-bearing bacterium. It can be found as part of the normal intestinal flora, and extensive contamination by *Cl. difficile* spores has been found in hospitals and care facilities for the elderly. Approximately 3 percent of healthy adults carry *Cl. difficile,* and half of all infants excrete it. Neither shows evidence of disease.

Both of the toxins produced by *Cl. difficile* are lethal. As little as 50 nanograms (ng) of either is sufficient to kill a mouse. Toxin A is cytotoxic (kills cells), lethal (kills animals), and an enterotoxin. Toxin B is a powerful enterotoxin causing both secretion of fluid and cell damage.

The disease caused by *Cl. difficile* is called pseudomembranous colitis, and is most likely to occur in hospitalized elderly patients who are receiving broad-spectrum antibiotics. Macrolide antibiotics such as clindamycin, lincomycin, or erythromycin are particularly likely to induce disease. It is thought that the antibiotics affect the normal flora, destroying the balance that holds *Cl. difficile* in check. It then overgrows, releases the toxins, and causes inflammation and fluid secretion from the colon.

CLINICAL FEATURES
Patients develop diarrhea of varying degrees of severity with more than four stools per day. These are watery and may contain blood. Often there are also cramping abdominal pains.

TREATMENT
Metronidazole is the antimicrobial of choice.

PREVENTION
No vaccine is available, but limiting the use of broad spectrum antibiotics is an important intervention. However, this is not always easy, especially with ill patients in hospital. It is estimated that at any one time, between 50 percent and 60 percent of patients in hospital will be on antibiotics.

corynebacterium diphtheriae/
diphtheria

Corynebacterium diphtheriae grows best in air enriched with carbon dioxide, and is the cause of diphtheria. It is rarely isolated from immunized populations but can be isolated from the throats of between 3 percent and 5 percent of unimmunized individuals. Disease results from the elaboration of its powerful toxin. A single molecule is lethal to most mammalian species.

The toxin is particularly lethal for cells in the cardiac, nervous, and renal systems. The bacterium is transmitted person-to-person by exchange of saliva or by the airborne route. It colonizes the throat and releases its toxin, causing destruction of the epithelial cells in the throat. This results in inflammation and the influx of defense cells, but the toxin kills these too. A grayish-green membrane of dead cells then forms over the whole area. The toxin also enters the bloodstream, targeting the heart, peripheral nerves, and kidneys. In tropical countries, *C. diphtheriae* can also cause wound infections.

CLINICAL FEATURES
Patients develop a sore throat and slight fever. The grayish-green membrane over the tonsils and pharynx is difficult to remove. Patients will become gravely ill with toxemia (toxin in the blood), and lose sensation and movement in the arms and legs (peripheral neuritis). Death occurs due to heart damage and failure.

TREATMENT
Antitoxin antibodies are given to neutralize existing toxin, then penicillin or erythromycin to kill the bacteria so that no more toxin is produced.

PREVENTION
A safe and effective vaccine consists of the inactivated diphtheria toxin (toxoid). It is a routine immunization given with tetanus toxoid and whooping cough vaccine to infants at two months of age and twice more thereafter. Boosters are required. This had led to the virtual elimination of diphtheria in Europe and the U.S.A.

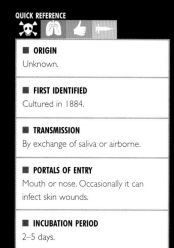

QUICK REFERENCE

■ **ORIGIN**
Unknown.

■ **FIRST IDENTIFIED**
Cultured in 1884.

■ **TRANSMISSION**
By exchange of saliva or airborne.

■ **PORTALS OF ENTRY**
Mouth or nose. Occasionally it can infect skin wounds.

■ **INCUBATION PERIOD**
2–5 days.

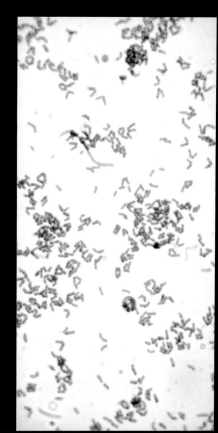

listeria monocytogenes/listeriosis

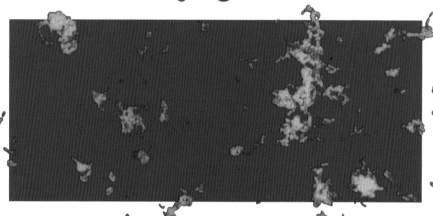

QUICK REFERENCE

■ **ORIGIN**
A ubiquitous environmental bacterium.

■ **FIRST IDENTIFIED**
The first isolation was in 1918 but it was not realized what the organism was until 1924.

■ **TRANSMISSION**
Acquired as a food poisoning.

■ **PORTAL OF ENTRY**
Gastrointestinal tract.

■ **INCUBATION PERIOD**
Up to 90 days.

Listeria monocytogenes can grow on simple bacteriological media such as blood agar, but also has the ability to invade, and multiply within, human and other cells. It is widely distributed in the environment, for example, in vegetation, dust, soil, and sewage, and can survive for long periods under optimal conditions.

It can contaminate food and is most often acquired by humans this way. It is particularly associated with unpasteurized milk or milk products, and soft cheeses. Food contamination after cooking can also be problematic. Infected adults rarely develop disease unless they are immune-suppressed, but listeriosis is particularly dangerous in newborn babies and during pregnancy. Bacteria enter the maternal blood-stream, cross the placenta, and infect the fetus.

CLINICAL FEATURES
Infection in pregnant women is asymptomatic in half of all cases. In the remainder there may be fever and a flu-like illness. Infection of the fetus results in intrauterine death, abortion, or delivery of an infected newborn. Newborns develop sepsis, meningitis, and pneumonia.

TREATMENT
Penicillin, ampicillin, and cotrimoxazole.

PREVENTION
Close attention to food hygiene.

bacillus cereus/food poisoning

Bacillus cereus is spore bearing, and its spores are widely distributed in the environment. In addition, it is a saprophyte growing in soil, which therefore contaminates most raw food and cereals (hence the name) including rice. It is relatively heat tolerant and can be found in enormous numbers in lightly cooked food.

The bacteria cause two forms of food poisoning. The first is vomiting, between one and six hours after consuming the contaminated food. For example, the *B. cereus* spores are present in rice, and are sufficiently heat-resistant to survive when the rice is boiled. During the gradual cooling of a large batch of boiled rice, for example in a take-out outlet, the spores germinate, producing vegetative (growing) bacteria that release an emetic toxin. Completely reheating the rice destroys the toxin. However, if the rice is only partially reheated, the toxin survives and is thus ingested by humans.

The second form of food poisoning caused by the *B. cereus* bacteria is diarrheal disease, which occurs between 8 and 16 hours after the contaminated food has been eaten. This time, the bacteria produce a different toxin that causes diarrhea. Infected patients need to eat a particular type of live bacteria that release their toxin in the intestine, as the *B. cereus* toxin is acid labile and is destroyed by the acid pH in the stomach.

CLINICAL FEATURES
Profuse vomiting or watery diarrhea with abdominal cramps and nausea.

TREATMENT
This is supportive. Antimicrobials are unnecessary as the food poisoning cures itself.

PREVENTION
No vaccine is available. Making sure that foods are properly cooked through is the mainstay of prevention.

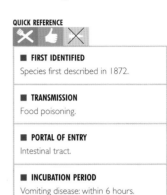

QUICK REFERENCE

■ **FIRST IDENTIFIED**
Species first described in 1872.

■ **TRANSMISSION**
Food poisoning.

■ **PORTAL OF ENTRY**
Intestinal tract.

■ **INCUBATION PERIOD**
Vomiting disease: within 6 hours.
Diarrheal disease: 8–16 hours.

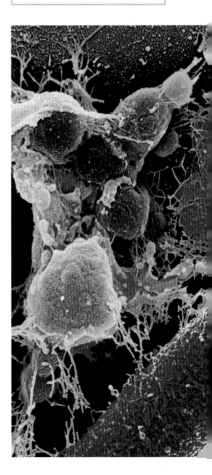

bacillus anthracis/anthrax

QUICK REFERENCE

■ **ORIGIN**
An ancient bacterium infecting animals.

■ **FIRST IDENTIFIED**
A disease of antiquity. *B. anthracis* was first isolated in 1896.

■ **TRANSMISSION**
Airborne spread of spores, inoculation of spores into skin, or ingestion of spores in meat.

■ **PORTALS OF ENTRY**
Respiratory tract, skin, or gastrointestinal tract.

■ **INCUBATION PERIOD**
Cutaneous anthrax: 2–3 days.
Intestinal anthrax: 5–24 days.
Inhalation anthrax: 6 days.

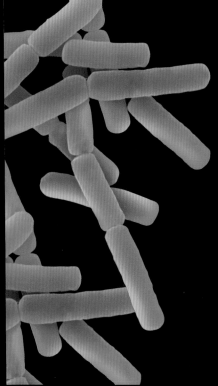

Anthrax is a disease of antiquity. Possibly the earliest reference to it came from the Bible where it is thought to have been the fifth (death of cattle) plague of Egypt in 1491 B.C. It is also claimed that the sixth plague of Egypt (cutaneous anthrax) was the first example of its use as a biological weapon: God told Moses to burn the cattle and sprinkle their ashes (which may have contained anthrax spores) in the air.

Bacillus anthracis spores are resistant to heat, drying, and chemicals, and when the British Army exploded a small bomb containing anthrax spores on Gruinard Island, off the Scottish coast, in 1941, the island remained contaminated for 45 years and was cleared only by drenching the island in seawater and formaldehyde. *Anthrax* is Greek for black, and the disease produces lesions on the skin that are covered by a black scab.

TRANSMISSION
Anthrax is a disease of domesticated and wild animals, who become infected through eating grass or soil containing anthrax spores. Humans become infected by contact with infected animals, their hides, or meat. Anthrax is not spread person-to-person. The disease has been eliminated from a number of countries but is still a significant problem in parts of Africa, in Greece, Turkey, Central Asia, and the Middle East.

DISEASE FORMS
B. anthracis possesses two toxins called edema factor, which causes tissue swelling, and lethal factor, which causes death. There are three major disease forms—cutaneous, intestinal, and inhalation anthrax—depending on the site of acquisition of the *B. anthracis* spores. Cutaneous anthrax occurs when spores are inoculated into minor abrasions in the skin. It is the most common form of anthrax and the least likely to result in death. Gastrointestinal anthrax occurs following ingestion of undercooked meat containing *B. anthracis* or its spores. Inhalation anthrax, or woolsorter's disease, occurs when spores derived from hides, wool, or fleeces are inhaled.

It has been estimated that release of 110 lb (50 kg) of weapon-grade anthrax spores from an aircraft over an urban population of five million would result in 250,000 cases of inhalation anthrax. In the U.S.A. in 2001, finely milled anthrax spores in envelopes were sent through the mail, resulting in a number of deaths and great public concern.

CLINICAL FEATURES

The initial lesion of cutaneous anthrax begins as a small pimple, two to three days after the spores have been inoculated into the skin. The pimple gradually enlarges and develops a vesicle on its surface before a ring of small vesicles emerge around it. These gradually acquire a bluish tinge and become dark and blood-stained. The central vesicle ulcerates and dries to form a dark-brown scab or "eschar." Over the next few days, this turns black. It is neither painful nor purulent, but there is a large amount of swelling (edema). In most cases, the lesion gradually heals but in about a fifth of cases, it disseminates to cause bacteremia, meningitis, and death.

Inhalation anthrax presents as a "flu-like" illness that lasts for three to four days. The patient then becomes severely ill, finding it difficult to breath, and with either a high fever or hypothermia. In sporadic cases, the mortality rate is between 80 percent and 90 percent. However, due to increased awareness and better diagnostic tools, in the recent terrorist outrage mentioned above the mortality rate was less than 40 percent.

TREATMENT

Penicillin, but "new" strains have been genetically modified and are now resistant.

PREVENTION

The live Sterne vaccine is used in animals. The human vaccine also contains this protective antigen but it has so far only been used in military personnel.

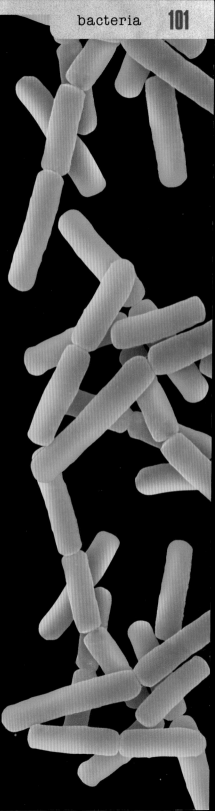

actinomyces israelii/
periodontal disease, actinomycosis

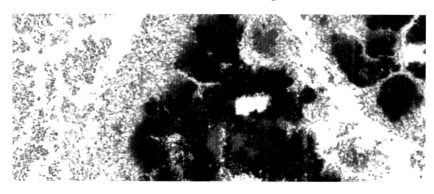

QUICK REFERENCE

■ **ORIGIN**
Part of the human bacterial flora.

■ **FIRST IDENTIFIED**
First human case described in 1878.

■ **TRANSMISSION**
An endogenous infection.

■ **PORTAL OF ENTRY**
Through breaches in the mucosa.

■ **INCUBATION PERIOD**
Unknown.

Actinomyces are a group of fungus-like bacteria. *A. israelii* is found in the human oral cavity, in the female genital tract, and in feces. It is a rare cause of human disease. However, in periodontal disease, it acts with a number of other bacteria to infect the gap between the teeth and gums, resulting in damage to enamel, and loosening and loss of teeth.

Actinomycosis is a chronic, pus-producing infection that is more common in males than females. It can affect the mouth and neck, abdomen, and chest, and in females, can cause pelvic infection.

CLINICAL FEATURES
A hard swelling of the skin overlies the site of infection. Often the infection bursts through the skin from the deeper abscess. The pus is thin but contains clumps of growing bacteria. Neck and mouth actinomycosis is usually derived from the teeth; abdominal actinomycosis is derived from a perforated appendix; and chest actinomycosis is derived from the spread of orofacial or abdominal infection. If untreated, it can prove fatal.

TREATMENT
High doses of penicillin over two to three weeks and, often, surgical treatment as well.

PREVENTION
No vaccine is available.

mycobacterium leprae/leprosy

The earliest mention of leprosy is in texts from 500 B.C. and although the disease is no longer in circulation in developed countries (it was eradicated from the U.K. in 1798), it is still present in 91 developing, mainly tropical countries, particularly Myanmar, India, and Indonesia. Fortunately, the number of cases is dropping— from 11 million in 1982 to 1.2 million in 1999.

Though leprosy can now be cured, the deformities caused prior to treatment still persist. *M. leprae* cannot be grown in artificial culture, and the armadillo is the only other animal that can be infected.

How leprosy is transmitted is not entirely clear, but it is probably from the nasal secretions of patients with the lepromatous leprosy form of the disease. Transmission seems to require prolonged contact. The targets for *M. leprae* are the cells that surround the nerves— the Schwann cells. The resulting disease is one of chronic and irreversible damage.

CLINICAL FEATURES

The incubation period is between three and five years. The disease usually begins with an indeterminate skin lesion that may fade. What happens next depends on the strength of the immune response to the bacterium. In tuberculoid leprosy, there are only a few skin lesions and some anesthesia due to nerve damage.

At the other end of the spectrum is lepromatous leprosy, whereby skin nodules containing large numbers of bacteria develop on cooler areas of the body, especially the ears. Coalescent nodules on the face give a leonine (lion-like) appearance, and if the cartilage of the nose is affected, can result in nasal collapse. The loss of sensation can lead to tissue damage by heat, cold, or wounding, and loss of digits or limbs.

TREATMENT

Combined dapsone, rifampicin, and clofazimine.

PREVENTION

The BCG vaccination.

QUICK REFERENCE

■ **ORIGIN**
From about 500 B.C. A description of a disease said to be leprosy in the Bible is probably another disease.

■ **FIRST IDENTIFIED**
First described in 1873.

■ **TRANSMISSION**
By nasal secretions.

■ **PORTAL OF ENTRY**
Probably the upper airways.

■ **INCUBATION PERIOD**
3–5 years.

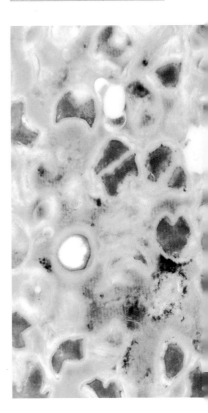

mycobacterium tuberculosis/
tuberculosis

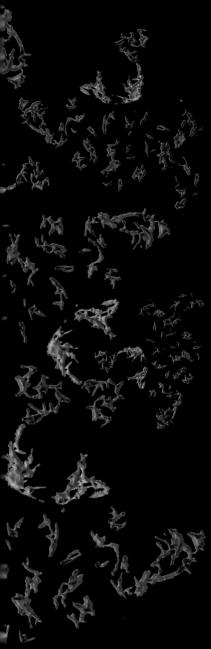

Tuberculosis is an old disease (skeletons from the Neolithic period and Egyptian mummies show signs of it), which once threatened to eliminate Western civilization and has been described as "the Captain of all the men of death" and the "White Plague." In England in the 1600s, around a fifth of all deaths were due to tuberculosis. Notables who died from the disease include the composers Purcell, Paganini, and Weber, and authors Beardsley, Camus, Molière, Kafka, and George Orwell.

In Europe, tuberculosis peaked in the late 18th and early 19th centuries, but with improved nutrition and the subsequent introduction of the Bacille Calmette Guerin (BCG) vaccine its impact was greatly diminished. Unfortunately, incidences of the disease in the developed world are now once again on the increase. However, tuberculosis has always been a major problem in developing countries, and has been made worse by the recent AIDS epidemic.

TRANSMISSION
The bacterium is transmitted person-to-person during coughing, when small droplets containing bacilli become airborne. These can remain suspended in the air; therefore, when patients cough out large numbers of bacteria in poorly ventilated spaces, transmission is easy. On average, one in six people in contact with a case of open tuberculosis becomes infected. It is also possible to get tuberculosis by ingestion, but this is usually due to *Mycobacterium bovis,* found in cow's milk.

In countries with a high prevalence of tuberculosis, children are the ones infected most frequently. Usually, there is just local replication of the bacterium in the upper part of the lung, with the immune system holding this in check so that no disease occurs. However, the *M. tuberculosis* is not eliminated and persists inside macrophages in the lung tissue and draining lymph nodes to reactivate later. In a minority of cases, the child's immune system does not hold the bacteria in check and they spread via the bloodstream to infect throughout the body, causing the patient to develop either tuberculous meningitis or disseminated tuberculosis (miliary TB).

In adults, tuberculosis occurs either due to reactivation of bacteria held quiescent since childhood by the immune system, or by new acquisition of *M. tuberculosis*.

CLINICAL FEATURES

Pulmonary tuberculosis is the most common form of *M. tuberculosis* infection. The primary infection in childhood is usually asymptomatic, but the post-primary, or reactivation, illness is characterized by fever, night sweats, aversion to food (anorexia) and subsequent weight loss, plus pulmonary signs of cough with mucoid or purulent sputum. The sputum may also be blood-stained (hemoptysis). At this stage, the patient will cough out large numbers of *M. tuberculosis* and is highly infectious.

Other manifestations include spinal tuberculosis, meningitis, abdominal tuberculosis, cutaneous tuberculosis (called lupus vulgaris because the patient has facial swelling giving a lupine, or wolf-like, appearance), and genitourinary tuberculosis.

TREATMENT

There were no specific drugs to treat tuberculosis until the development of streptomycin in 1948. Unfortunately, the bacteria soon developed resistance to this treatment, and this has led to the current strategy of combining three or four agents, including isoniazid, streptomycin, pyrinzinamide, and rifampicin. This works well if patients adhere to a three- to six-month therapeutic regimen. However, a number of drug-resistant strains have recently emerged, raising the specter of untreatable tuberculosis.

PREVENTION

The live attenuated BCG vaccine is not ideal but is currently all that is available.

QUICK REFERENCE

■ **ORIGIN**
An ancient disease referred to in Chinese literature around 4,500 years ago.

■ **FIRST IDENTIFIED**
Aristotle and Galen recognized that the disease was infectious.

■ **TRANSMISSION**
Airborne via coughing, through ingestion or, occasionally, as a result of direct contact.

■ **PORTALS OF ENTRY**
Lungs and intestinal tract or, less often, through the skin.

■ **INCUBATION PERIOD**
Not applicable.

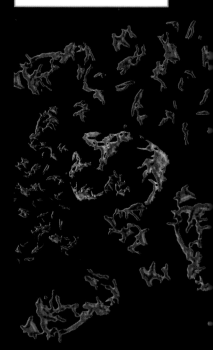

mycobacterium marinum/
fish-tank granuloma

QUICK REFERENCE

- **ORIGIN**
A pathogen of fish.

- **FIRST IDENTIFIED**
In the 1940s.

- **TRANSMISSION**
In water, originally from fish, but can survive for long periods in water.

- **PORTALS OF ENTRY**
Through skin abrasions and cuts.

- **INCUBATION PERIOD**
Days to weeks.

Mycobacterium marinum is a nontuberculous mycobacterium that causes only mild, limited disease but can produce more severe infections in those with defective immune systems. *M. marinum* is primarily a pathogen of fish, and infection (glanuloma) in humans is associated with improperly maintained swimming pools or, more frequently, with cleaning and maintaining fish tanks. The bacterium is introduced through minor cuts or abrasions in the skin.

CLINICAL FEATURES
Swimming pool granulomas usually occur on the knees, elbows, or hands, though improved swimming-pool construction and chlorination means they are now less common. Fish-tank granulomas usually occur on the hands or wrists. They appear as warty inflamed nodules with surrounding swelling and crust formation. Occasionally they spread locally on the skin or colonize lymphatics.

TREATMENT
Lesions often resolve without treatment but, if necessary, a combination of rifampicin and ethambutol is very successful.

PREVENTION
Prevention is by properly maintaining swimming pools and wearing gloves or covering cuts with waterproof dressings when maintaining or cleaning fish tanks.

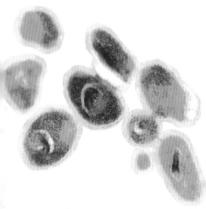

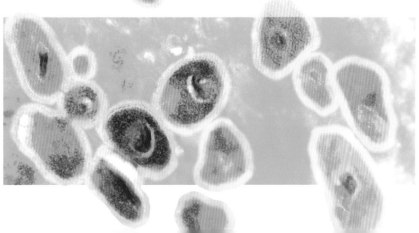

mycobacterium avium and intracellulare/lymphadenitis

Mycobacterium avium and *M. intracellulare,* together with *M. scrofulaceum,* form a complex of mycobacteria called the *M. avium* complex, or MAC. They are widely distributed in the environment. In the southern U.S.A., they are found in brown-water swamps but can also be isolated from soil dust, a variety of animal species, and tap water worldwide.

Infection in children, usually under five years of age, gives enlargement of a lymph gland in the neck (cervical adenitis), usually the lymph node draining the tonsils. In AIDS patients, whose immune systems are compromised, the bacteria are disseminated around the body, usually from the intestine. *M. avium* and *M. scrofulaceum* are more frequently associated with cervical adenitis, and *M. intracellulare* with AIDS-associated disseminated disease.

CLINICAL FEATURES
There is usually considerable enlargement of a single lymph node in the neck. If left untreated, pus-filled node points can burst through to the skin leaving a chronic discharging sinus. This process can take years. In the disseminated disease, patients are wasted and febrile with abdominal pain, night sweats, and diarrhea.

TREATMENT
Surgical excision can cure cervical adenitis, but must be carried out carefully as there are important nerves in this region that must not be damaged. In addition, if the node leaks or bursts during surgery its contents will reinfect the region. Clarithromycin has been used as an adjunct to surgery. In cases of disseminated disease, antiretroviral drugs are given to restore the immune system.

PREVENTION
No vaccine is available.

QUICK REFERENCE

■ **ORIGIN**
Widely distributed environmental bacteria.

■ **FIRST IDENTIFIED**
M. avium was first isolated in 1890.

■ **TRANSMISSION**
From the environment.

■ **PORTAL OF ENTRY**
Mouth.

■ **INCUBATION PERIOD**
Unknown.

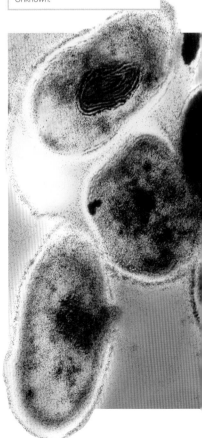

tropheryma whipplei/
whipple's disease

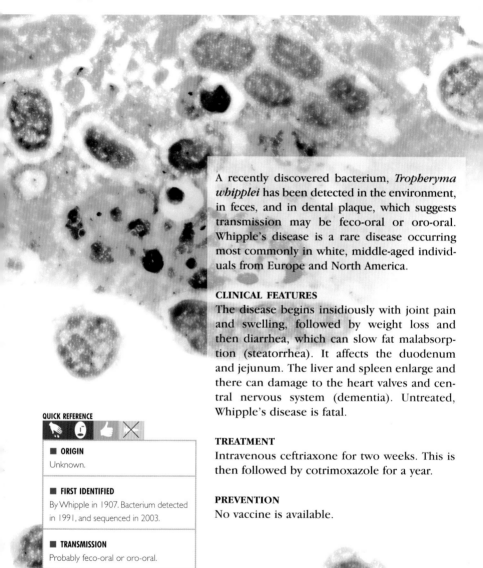

A recently discovered bacterium, *Tropheryma whipplei* has been detected in the environment, in feces, and in dental plaque, which suggests transmission may be feco-oral or oro-oral. Whipple's disease is a rare disease occurring most commonly in white, middle-aged individuals from Europe and North America.

CLINICAL FEATURES
The disease begins insidiously with joint pain and swelling, followed by weight loss and then diarrhea, which can slow fat malabsorption (steatorrhea). It affects the duodenum and jejunum. The liver and spleen enlarge and there can damage to the heart valves and central nervous system (dementia). Untreated, Whipple's disease is fatal.

TREATMENT
Intravenous ceftriaxone for two weeks. This is then followed by cotrimoxazole for a year.

PREVENTION
No vaccine is available.

QUICK REFERENCE

■ **ORIGIN**
Unknown.

■ **FIRST IDENTIFIED**
By Whipple in 1907. Bacterium detected in 1991, and sequenced in 2003.

■ **TRANSMISSION**
Probably feco-oral or oro-oral.

■ **PORTAL OF ENTRY**
Intestinal tract.

■ **INCUBATION PERIOD**
Unknown.

chlamydia psittaci/psittacosis

Chlamydia psittaci is an obligate intracellular bacterium. It causes infection in birds, usually asymptomatically, and is transmitted accidentally to humans. There is little evidence of human-to-human transmission. Although its name implies that it infects psittacine birds such as parrots and budgerigars, it has a much wider range, including pigeons, ducks, and passerine birds. It is excreted by these birds in their feces. Human infection is via inhalation, generally in adults between the 30 and 60 years of age.

CLINICAL FEATURES
Patients present with a flu-like illness with fever, malaise, aversion to food (anorexia), rigors, sore throat, headache, and photophobia (extreme sensitivity to light). This may worsen to give severe bronchopneumonia, delirium, liver and spleen enlargement, and even endocarditis. It is a chronic infection which can result in death.

TREATMENT
Macrolides such as azithromycin and clarithromycin, or tetracycline, are effective.

PREVENTION
No vaccine is available.

QUICK REFERENCE

■ **ORIGIN**
A widely distributed bacterium of birds.

■ **FIRST IDENTIFIED**
C. psittaci was first cultured in 1930.

■ **TRANSMISSION**
Airborne, from dried bird droppings, via inhalation.

■ **PORTAL OF ENTRY**
Respiratory tract.

■ **INCUBATION PERIOD**
Around 10 days.

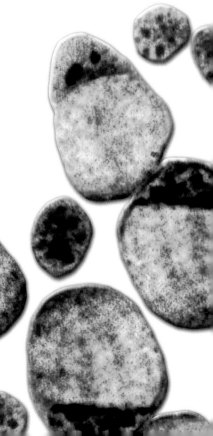

chlamydia trachomatis/
sexually transmitted infection, trachoma

Chlamydia trachomatis is subdivided into a number of serotypes. Serotypes A–C cause trachoma in tropical countries. This is a chronic scarring disease of the eyes and the most common infective cause of blindness worldwide. Serotypes D–K are transmitted sexually, but can also cause conjunctivitis and, later, pneumonia in babies as they pass through an infected birth canal. Serotypes L1–L3 cause the sexually transmitted disease lymphogranuloma venereum.

CLINICAL FEATURES
The initial phase of trachoma is conjunctivitis, which leaves small pits in the conjunctivae. This heals by scarring but leads to distortion of the eyelids, which cannot close properly and invert. This causes the eyelashes to rub on the eyeball, leading to further scarring and then blindness.

The sexually transmitted infections are urethritis in males, which may extend to the testicles, causing painful swelling, and urethritis and cervicitis in females. The latter may ascend to the uterus to infect the fallopian tubes. This often leads to blockage of the tubes and infertility.

Lymphogranuloma venereum is a destructive tropical disease that moves to the inguinal lymph nodes draining the lymphatics of the genitalia. These enlarge, rupture, and drain by sinuses to the skin. There will also be ulcers, granulomas, then elephantiasis of the genitalia.

TREATMENT
Trachoma can be treated with tetracycline or erythromycin in its acute phase. Chlamydial sexually transmitted infections are treated with azithromycin or tetracycline. Neonatal conjunctivitis is treated with erythromycin or azithromycin orally.

PREVENTION
No vaccines are available. Wearing condoms can prevent transmission of *C. trachomatis* D–K and L1, L2, and L3.

chlamydia pneumoniae/
respiratory tract infection

Chlamydia pneumoniae was first cultivated in Taiwan in 1976 and called the TWAR (Taiwan acute respiratory) agent. It is now clear that it has a worldwide distribution and is an important cause of acute respiratory tract infection particularly in children and young adults. Severe disease is most likely to occur in elderly patients or those with preexisting lung or heart damage. It is estimated that between 60 percent and 80 percent of the world's population has been infected.

There is evidence that following initial infection the bacteria can persist, perhaps in white blood cells, for years, before reactivating. The bacteria has also recently been linked with coronary artery disease.

QUICK REFERENCE

- **ORIGIN**
  A bacterium of humans.

- **FIRST IDENTIFIED**
  First isolated in 1976.

- **TRANSMISSION**
  Airborne via respiratory secretions, or from hand to mouth.

- **PORTAL OF ENTRY**
  Upper respiratory tract.

- **INCUBATION PERIOD**
  Around 21 days.

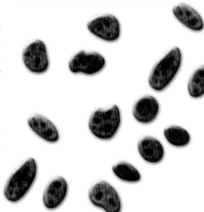

CLINICAL FEATURES
Patients develop fever and an upper respiratory tract infection. However, disease may progress to lower tract infection with pneumonia. In some cases, it causes sinusitis or middle-ear infection (otitis media).

TREATMENT
Tetracycline or macrolides such as erythromycin, azithromycin, or clarithromycin are suitable antimicrobials.

PREVENTION
No vaccine is available.

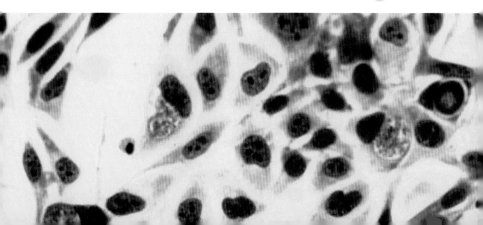

coxiella burnetii/Q-fever

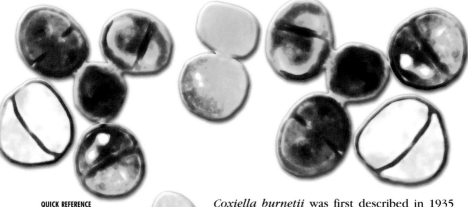

■ **ORIGIN**
A bacterium of animals, transmitted animal-to-animal by ticks.

■ **FIRST IDENTIFIED**
Q-fever was first described in 1935.

■ **TRANSMISSION**
Airborne or, occasionally, by ingestion.

■ **PORTALS OF ENTRY**
Respiratory tract or intestines.

■ **INCUBATION PERIOD**
2 weeks.

Coxiella burnetii was first described in 1935 after an outbreak of typhoid-like fever among abattoir workers in Queensland, Australia. Q-fever has a worldwide distribution but most cases are reported from the U.K. and France, though it is likely that this is a result of under-recognition in other countries.

Airborne transmission of the bacterium is from infected animal tissues or products. Occasionally, transmission occurs as a result of ingestion of unpasteurized milk. Ticks appear to be biological vectors of transmission between animals but not to humans. There is little evidence of onward person-to-person transmission. Q-fever is an occupational illness and those most often infected are abattoir workers, farmers, and vets. Rarely, it infects the placenta and fetus, leading to abortion.

CLINICAL FEATURES
Patients develop a fever with headache, chills, muscle and joint pains, photophobia (extreme sensitivity to light), pharyngitis, and diarrhea. However, many infections are subclinical. If the patient has damaged heart valves, a progressive and fatal endocarditis can occur.

TREATMENT
Tetracycline antibiotics.

PREVENTION
No vaccine is available.

rickettsia typhi and r. prowazekii/
typhus

Rickettsia typhi and *R. prowazekii* bacteria have animal hosts, and humans are accidentally infected when bitten by an insect (a tick or flea) that has previously fed on the reservoir host.

R. typhi causes flea-borne, or murine, typhus. It has a worldwide distribution but is mainly found in the tropics and subtropics and is associated with poverty and overcrowding. The main hosts for *R. typhi* are the black rat and the brown rat, and infection is transmitted by the rat flea.

R. prowazekii causes epidemic, or louse-borne, typhus. It is transmitted among humans by the body louse and epidemics occur in refugee camps, prisons, and at times of war or natural disaster. The flying squirrel appears to be the natural host.

QUICK REFERENCE

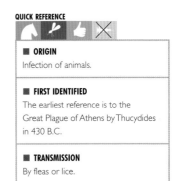

■ **ORIGIN**
Infection of animals.

■ **FIRST IDENTIFIED**
The earliest reference is to the Great Plague of Athens by Thucydides in 430 B.C.

■ **TRANSMISSION**
By fleas or lice.

■ **PORTAL OF ENTRY**
Through the skin.

■ **INCUBATION PERIOD**
6–15 days.

CLINICAL FEATURES
Louse-borne typhus can be a severe disease. Patients develop headache and fever. Over the next few days, they may develop a maculo-papular rash. In severe disease, the rash becomes hemorrhagic and patients develop myocarditis, damage to the central nervous system, and renal damage. The bacterium can remain latent in the patient and reactivate later to produce a similar, but milder, disease (Brill-Zinsser disease).

Flea-borne typhus is similar to epidemic louse-borne typhus but is less severe and does not reactivate.

TREATMENT
Tetracycline antibiotics.

PREVENTION
No vaccine is available but avoiding fleas and lice prevents infection.

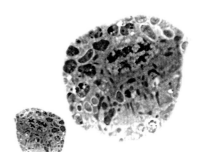

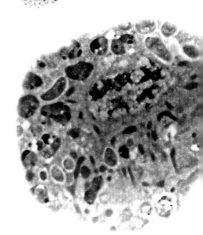

ehrlichia chaffeensis/
monocytic ehrlichiosis

QUICK REFERENCE

■ **ORIGIN**
A bacterium of deer.

■ **FIRST IDENTIFIED**
In Ford Chaffee, U.S.A., in 1986.

■ **TRANSMISSION**
Ticks are the biological vectors.

■ **PORTAL OF ENTRY**
Through the skin by tick bite.

■ **INCUBATION PERIOD**
5–6 days.

In the U.S.A., *Ehrlichia chaffeensis* is transmitted between deer and to humans by ticks. Monocytic ehrlichiosis is a newly described disease. The first cases were detected in Fort Chaffee, Arizona, in 1986. Cases have since been detected in Thailand and Africa but it is unclear what animal is the reservoir host and which tick is the vector. Often, it is adult males who are infected, particularly hunters and golfers.

CLINICAL FEATURES
Patients present with a high fever and 10 percent will have a fine maculopapular rash. There may also be evidence of the tick bite as a black scab or "eschar." In about 15 percent of patients, the disease progresses to become life-threatening, with septic shock, respiratory distress, and loss of nerve function.

TREATMENT
Tetracyclines such as doxicycline are of benefit.

PREVENTION
No vaccine is available, but avoidance of ticks prevents infection.

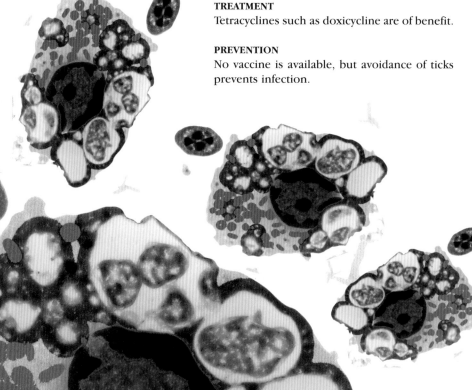

anaplasma phagocytophilum/
human granulocytic ehrlichiosis

So far, human granulocytic ehrlichiosis disease has only been recognized in the U.S.A. and Europe. It is likely that small rodents are the reservoir host and humans become infected through the bite of hard ticks.

CLINICAL FEATURES

Patients develop a fever with headache, muscle and joint pains, rigors, and nausea. Those who are severely affected (around a quarter of cases) develop respiratory problems that require ventilatory support and opportunist infections similar to those that affect AIDS patients.

TREATMENT

Tetracyclines such as doxycycline.

PREVENTION

Avoiding contact with ticks.

QUICK REFERENCE

■ **ORIGIN**
A bacterium of animals.

■ **FIRST IDENTIFIED**
The first human cases were diagnosed in 1994.

■ **TRANSMISSION**
By the bite of ixodid ticks.

■ **PORTAL OF ENTRY**
Through the skin to the bloodstream.

■ **INCUBATION PERIOD**
1–60 days after the tick bite.

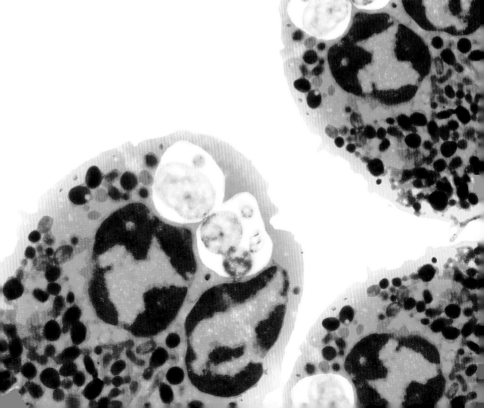

bartonella bacilliformis/
oroya fever: carrion's disease

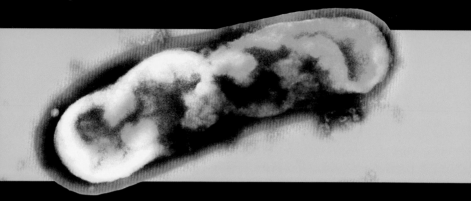

■ **ORIGIN**
Unknown.

■ **FIRST IDENTIFIED**
Oroya fever was first described in 1870.

■ **TRANSMISSION**
Sandfly bites.

■ **PORTAL OF ENTRY**
Through the skin to the bloodstream.

■ **INCUBATION PERIOD**
Around 20 days.

B. bacilliformis is responsible for outbreaks of a severe and frequently fatal infection of humans in the mountainous regions of Peru, Colombia, and Ecuador. The infection is transmitted by the sandfly. It appears that humans are the only vertebrate host of *B. bacilliformis*. The bacteria can persist in their human host for long periods after recovery from the Oroya fever.

CLINICAL FEATURES
Patients develop a high fever. *B. bacilliformis* destroys red and other blood cells so patients become anemic. There is also enlargement of the liver and spleen, and hemorrhage into lymph nodes. If untreated, the mortality rate is more than 40 percent.

Skin eruptions develop up to 10 weeks after recovery from Oroya fever. These consist of round, elevated hard nodules on the legs, arms, and face, and may persist for more than a year. Around half of those who become infected with *B. bacilliformis* remain asymptomatic but may subsequently develop skin eruptions.

TREATMENT
Chloramphenicol, though clarithromycin may be more effective.

PREVENTION
No vaccine is available. Avoiding sandflies, which feed at night, is the best prevention.

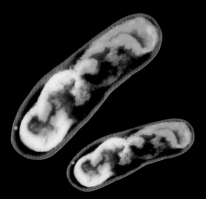

bartonella henselae/
cat-scratch disease

Bartonella henselae is the cause of cat-scratch disease. Cats can be persistently infected with the bacterium, which is transmitted to humans from a cat scratch or bite. The bacterium gets into the skin and tracks to the local lymph nodes. These enlarge and the patient becomes febrile. In patients with AIDS or other forms of immunosuppression, the bacterium can disseminate to cause bacillary angiomatosis or bacillary peliosis which affect the liver and spleen.

    B. clarridgeiae causes an identical disease but appears to be transmitted by cat fleas. In the U.S.A. it is estimated that there are around 24,000 cases each year, 2,000 of which result in hospitalization.

CLINICAL FEATURES
Patients develop fever and enlarged pus-filled lymph nodes which drain the scratch or bite area.

TREATMENT
Antimicrobials such as tetracycline, gentamicin, chloramphenicol, or clarithromycin are all effective. Treatment must be continued for six weeks.

PREVENTION
No vaccine is available.

QUICK REFERENCE

■ **ORIGIN**
Unknown.

■ **FIRST IDENTIFIED**
The bacterium was detected by molecular technology in 1990, but the earliest description of the disease was in 1889.

■ **TRANSMISSION**
By cat scratch or bite, or cat fleas.

■ **PORTAL OF ENTRY**
Through the skin.

■ **INCUBATION PERIOD**
Weeks to months.

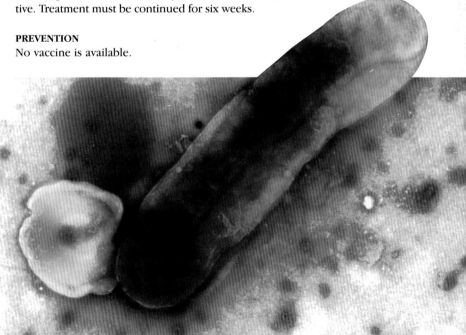

mycoplasma pneumoniae/
respiratory tract infection

■ **ORIGIN**
A solely human pathogen.

■ **FIRST IDENTIFIED**
First grown in 1962.

■ **TRANSMISSION**
By respiratory droplets.

■ **PORTAL OF ENTRY**
Upper airways.

■ **INCUBATION PERIOD**
6–23 days.

This bacteria is the only mycoplasma that is a major pathogen for humans. It is a mollicute (from the Latin mollis and cutis meaning soft skin), meaning that it does not have a cell wall. This has the effect of making the bacterium resistant to those antibiotics which inhibit cell wall synthesis such as penicillins, cephalosporins, and vancomycin.

M. pneumoniae is solely a human pathogen although other animals can be infected experimentally. It is endemic in human populations and produces epidemics every five years or so. It can only be transmitted through close contact. Though *M. pneumoniae* is not usually part of the normal flora of the oropharynx, it can persist for some time after infection. Most cases occur in winter. It mainly attacks children of about five years of age and young adults.

CLINICAL FEATURES
There is a gradual onset of headache, fever, and a sore throat. This is followed by coughing fits. If the disease progresses, there will be chest pain, especially when coughing. The disease can also be a cause of middle-ear infection (otitis media).

TREATMENT
Macrolides such as erythromycin, azithromycin, or clarithromycin are effective, but only if taken over several days.

PREVENTION
No vaccine is available.

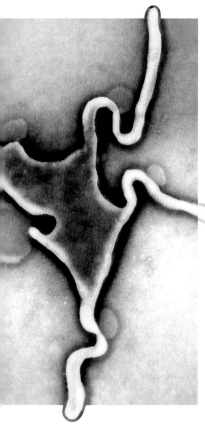

neisseria gonorrhoeae/gonorrhea

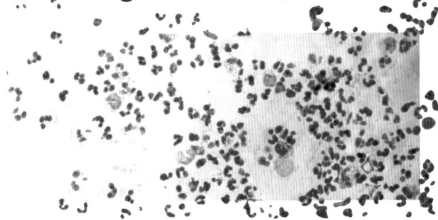

Neisseria gonorrhoeae is the cause of the sexually transmitted infection gonorrhea, which requires close intimate contact in order to spread. Gonorrhea was well controlled until the 1960s when changes in social mores, oral contraception, frequent travel, and migration led to an enormous increase in the number of cases.

The figures dropped in the early 1990s, perhaps due to the increased practice of safe sex and with the introduction of a potent antibiotic that was effective even in a single dose. However, resistance has developed, people have become complacent, and gonorrhea is again on the up.

CLINICAL FEATURES

In males, gonorrhea is characterized by a purulent discharge from the penis and pain on passing urine. It can also cause infection of the rectum or a sore throat. In females it causes urethritis and cervicitis, which presents as a vaginal discharge. Females can also be infected asymptomatically. If born through an infected cervix, babies are at risk of developing a very severe conjunctivitis. Untreated, this leads to loss of the eye(s).

TREATMENT

There are very few antibiotics that can be relied on to be effective without testing them in vitro.

PREVENTION

Using a condom decreases the risk of infection.

QUICK REFERENCE

■ **ORIGIN**
Solely a human pathogen that has evolved from the meningococcus some 1,500–2,000 years ago.

■ **FIRST IDENTIFIED**
Bacterium was first cultured in 1936.

■ **TRANSMISSION**
By close contact and the exchange of genital secretions.

■ **PORTALS OF ENTRY**
The mucosa of the genitalia, rectum, or throat.

■ **INCUBATION PERIOD**
From 3 days to several weeks.

neisseria meningitidis/
meningococcal disease

Neisseria meningitidis, or the meningococcus, is solely a human pathogen. Meningococci are part of the normal flora of the oropharynx and as many as half of us have the bacteria residing in the tonsils or on the mucosa. Once established, they are probably of benefit to our immune systems; thus, it is only when a new virulent strain is acquired that disease results.

Meningococcal disease is one of the few acutely life-threatening infections still remaining in developed countries. The incidence of disease in the U.K. is small, and with improvements in patient management, early recognition, and improved intensive care, the mortality rate for severe disease has fallen from 76 percent in 1986 to under 3 percent in 2003. Most cases occur in those under a year old.

INFECTION GROUPS

In temperate countries, Group B and C infections occur all year around but with major upsurges in winter, though these are generally sporadic, meaning that they are not as widely spread as an epidemic.

However, in Africa, where the Group A meningitis belt stretches from Senegal and Gambia in the west to Ethiopia in the east, the disease is characterized by epidemics that occur every three, four, or five years (though not in each country at the same time). There is now also evidence that the belt has extended northwest to Tunisia and south to Zambia, Malawi, Namibia, and Angola. The epidemics occur in the hot dry season and usually cease with the onset of the rains. They predominantly affect children and adolescents but as the epidemic gains hold, the average age of patients rises.

In both developed and developing countries, transmission is by exchange of upper respiratory tract secretions by coughing or exchange of saliva, and requires close contact. The virulent meningococcus then colonizes the oropharynx and in a minority of cases, gets into the bloodstream. Here, the meningococci multiply to reach enormous concentrations (greater than the concentration of blood cells in our bodies), causing massive, uncontrollable activation of inflammatory molecules called cytokines and chemokines, which precipitate disease.

CLINICAL FEATURES

Although meningococci can cause arthritis, conjunctivitis, or pneumonia, they more frequently manifest themselves as the more major meningococcal meningitis (MM), meningococcal septicemia (MS), or a combination of the two (MM and MS).

MS presents with a rapid onset of a hemorrhagic rash that can be small purple spots that do not vanish when pressed (petechiae), larger bruise-like lesions (purpura), or even bigger areas of bleeding into the skin of whole limbs (ecchymosis). In each case, it results from damage to the small blood vessels (capillaries), which then leak blood into the skin and other areas. The rash and illness progress rapidly, causing shock and, if untreated, death. Meningitis alone is much less likely to result in death. It presents with fever, neck stiffness, headache, and photophobia (extreme sensitivity to light).

TREATMENT

Early treatment with injected penicillin or a cephalosporin such as ceftriaxone is of vital importance. Parents and anyone responsible for children must be able to recognize the rash so that treatment can begin as soon as possible. The severity of the disease should be assessed and, if the patient is severely ill, he or she should be admitted to an intensive care unit.

PREVENTION

The conjugate Group C vaccine has almost eliminated disease in the U.K. However, there is no vaccine for the Group B meningococcus. A polysaccharide Group A vaccine is available in the African meningitis belt.

QUICK REFERENCE

■ **ORIGIN**
A human bacterium.

■ **FIRST IDENTIFIED**
The disease was first described in 1806 and bacterium identified in 1887.

■ **TRANSMISSION**
By close contact through saliva or coughing.

■ **PORTAL OF ENTRY**
Oropharynx.

■ **INCUBATION PERIOD**
Indefinite, but probably less than a week.

moraxella catarrhalis/
respiratory tract infection

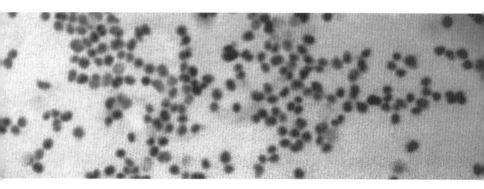

QUICK REFERENCE

■ ORIGIN
Part of the normal flora of the upper airways.

■ FIRST IDENTIFIED
In 1896.

■ TRANSMISSION
An endogenous infection derived from the patient's normal flora.

■ PORTAL OF ENTRY
Not applicable.

■ INCUBATION PERIOD
Not applicable.

The human pathogen *Moraxella catarrhalis* is part of the normal flora of the mouth, throat, and larynx. It is principally an opportunist pathogen, which means it causes infection when there is damage to immunity.

The infections it causes include exacerbation of chronic obstructive lung disease (previously referred to as chronic bronchitis) and pneumonia in patients with preexisting lung damage (for example from working in the coal, steel, and chemical industries). It can also cause sinusitis and middle-ear infection (otitis media).

CLINICAL FEATURES
The most frequent manifestation is exacerbation of chronic bronchitis. Chronic bronchitis is a response of the lung to long-term damage by cigarette smoking, air pollution, or smoke. It is characterized by changes in the lining of the bronchi (airways) and excessive production of mucus (phlegm). This can lead to blockage and bacteria becoming trapped in the lungs so that a low-grade infection occurs, characterized by fever, difficulty breathing, and the production of purulent sputum.

TREATMENT
Cotrimoxazole and ciprofloxacin may be useful.

PREVENTION
No vaccine is available.

haemophilus influenzae/
meningitis, pneumonia

Haemophilus influenzae was originally cultured by Pfeiffer in 1887 when he thought he had discovered the cause of influenza. There are six recognized strains, and of these, *H. influenzae*(b) is the most important cause of disease in humans.

H. influenzae and *H. influenzae*(b) are part of our normal flora, and are mostly found in the naso-oropharynx. *H. influenzae*(b) infections have been virtually eradicated in countries that use the Hib conjugate vaccine. Elsewhere (and prior to the introduction of the Hib vaccine), it caused infections, mainly in children under five, including pneumonia, meningitis, septic arthritis, epiglottitis, and osteomyelitis. In unvaccinated communities, rates of between 26 and 1,100 cases per 100,000 have been found.

Invasive *H. influenzae*(b) disease is more common in the winter months. Disease occurs when the bacteria move from their colonization sites in the upper respiratory tract to the bloodstream or into the lungs.

CLINICAL FEATURES

Epiglottitis is a life-threatening infection. The epiglottis acts like a flap-valve to prevent food and drink from going down the larynx to the lungs. If it becomes very inflamed and swollen, this can block breathing, causing instant death.

TREATMENT

Ceftriaxone has had an effect on all isolates.

PREVENTION

The Hib conjugate vaccine (polysaccharide chemically linked to protein) works extremely well, even in infants.

QUICK REFERENCE

■ **ORIGIN**
Part of our own bacterial flora.

■ **FIRST IDENTIFIED**
First viewed microscopically in 1891 but first isolation in 1887.

■ **TRANSMISSION**
Person-to-person by the respiratory route to become normal flora.

■ **PORTAL OF ENTRY**
From upper airways mucosa to the bloodstream or lungs.

■ **INCUBATION PERIOD**
Not applicable.

bordetella pertussis/whooping cough

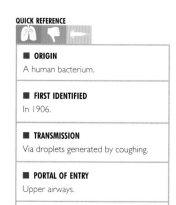

- **ORIGIN**
A human bacterium.

- **FIRST IDENTIFIED**
In 1906.

- **TRANSMISSION**
Via droplets generated by coughing.

- **PORTAL OF ENTRY**
Upper airways.

- **INCUBATION PERIOD**
5–21 days.

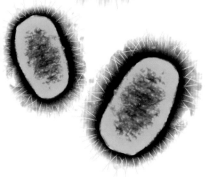

There is no known animal or environmental reservoir of *B. pertussis* but its close relative, *B. parapertussis,* causes disease in sheep. It appears that it is maintained in the human population by carriers, who do not necessarily have disease.

In unvaccinated countries, whooping cough is a major cause of damage and death. Most cases occur in children under 15 years of age, and the younger they are, the more severe the disease. We are now seeing cases in middle-aged adults whose immunity has waned since vaccination, but in these patients, the disease is usually less severe.

The bacterium is transmitted person to person via respiratory secretion during coughing. *B. pertussis* adheres to the linings of the airways and releases a vast array of toxins that lead to fever and the characteristic whooping cough.

CLINICAL FEATURES
The disease occurs in three phases. The first is the coryzal phase in which the patient has a runny nose, fever, and a dry cough. At this stage, the patient excretes large numbers of *B. pertussis* and is infectious. The disease worsens over the following two weeks until paroxysmal coughing begins. This phase lasts for three weeks and consists of long bursts of coughing followed by a short inspiratory "whoop" as the child tries to draw in air as quickly as possible before the next paroxysm.

During this time, damage may include bleeding into the conjunctivae, bursting or collapse of the lungs, and damage to the tongue caused by biting. There are convulsions, and occasionally brain damage. The patients then enters the convalescent phase with occasional paroxysms, which can last for more than six months.

TREATMENT
This is largely supportive. Antibiotics such as erythromycin are given to decrease the infectiousness of the patient.

PREVENTION
Two safe and effective vaccines are available.

brucella melitensis/brucellosis

Brucella melitensis is named after the Roman name for Malta (Melita, the honey island). Its close relatives *B. abortus*, *B. suis*, and *B. canis* are named so because they cause abortion in cattle (*abortus*), and infect pigs (*suis*), or dogs (*canis*). Brucellosis is particularly found in Mediterranean countries, the Arabian peninsula, Indian subcontinent, and parts of Central and South America. However, there has been a recent outbreak in Northern Ireland.

B. melitensis is particularly found in sheep and goats. The bacteria chronically infect their host species and are excreted throughout their lifetimes in urine, feces, milk, and products of conception. They are transmitted to humans through cuts and abrasions in the skin or mucosa, or by ingestion of unpasteurized milk. Person-to-person spread is rare.

After entering the human body, the bacteria travel through the lymphatics to the local lymph nodes where they grow and are eventually shed into the bloodstream.

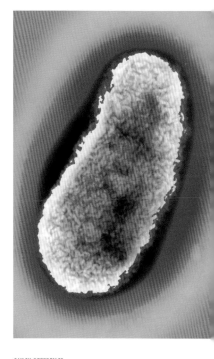

CLINICAL FEATURES
The infection can present in a large variety of ways, but fever and enlarged lymph nodes are the most common. The fever waxes and wanes, giving the disease its alternative name, undulant fever. Other features include abdominal pain, dry cough, arthritis, endocarditis, testicular inflammation, or even psychiatric problems. The infection can progress to chronic brucellosis where some of the features persist for years. Only around 2 percent of untreated cases prove fatal. However, the disease is very debilitating.

TREATMENT
A combination of rifampicin or gentamicin with doxycycline.

PREVENTION
There is no human vaccine, but a live-attenuated vaccine is used in cattle and other animals to try to eliminate it from the reservoir hosts.

QUICK REFERENCE

■ **ORIGIN**
A pathogen of animals accidentally transmitted to humans.

■ **FIRST IDENTIFIED**
The disease was first described in 1856. *B. melitensis* was first isolated in 1886, *B. abortus* in 1895, *B. suis* in 1914, and *B. canis* in 1966.

■ **TRANSMISSION**
From animals in meat, meat products, milk; or placenta and fetus.

■ **PORTAL OF ENTRY**
Skin or mucous membranes of the respiratory tract or intestine.

■ **INCUBATION PERIOD**
2–3 weeks, but can be up to 2 months.

legionella pneumophila/
legionnaire's disease

QUICK REFERENCE

■ **ORIGIN**
An environmental bacterium.

■ **FIRST IDENTIFIED**
The first case of Legionnaire's disease was found, retrospectively, in 1954 but a major outbreak occurred in 1976. The bacterium was isolated in 1977.

■ **TRANSMISSION**
By small airborne droplets and inhalation.

■ **PORTAL OF ENTRY**
Lower respiratory tract.

■ **INCUBATION PERIOD**
Pontiac fever: 24–47 hours.
Legionnaire's disease: 2–10 days.

A dramatic outbreak of severe respiratory tract infection occurred at an American Legion convention in Philadelphia, U.S.A., in 1976, and out of the 4,400 Legionnaires, 182 became infected and 29 died. This led to the search for the causative agent of the infection now known as Legionnaire's disease.

Eventually, a bacterium grown in fertile hens' eggs was identified and given the name *Legionella pneumophila*, and it soon became apparent that the newly isolated bacterium had been responsible for other outbreaks of the disease in the U.S.A., in 1965 in Washington, D.C., in 1968 in Pontiac, and in 1974 in the same Philadelphia hotel. In each case, there was an apparent link with air-conditioning systems.

TRANSMISSION
We now know that *L. pneumophila* is an environmental bacterium found in river mud, streams, and lakes, where it can survive relatively well in water temperatures up to 120°F (50°C). In addition, it is quite resistant to water chlorination. The bacteria are able to grow in water that is almost devoid of nutrients, but particularly like the materials used in plumbing, such as rubber washers, pipe sealants, and shower heads.

They appear to have evolved their virulence mechanism to avoid being destroyed by the amebae found in watery environments which normally engulf and kill bacteria to obtain nutrients. To do this, the amebae use mechanisms similar to those used by our own major lung defense cells (macrophages). Thus, if *L. pneumophila* bacteria accidentally gain access to our lower airways they are able to survive and cause pneumonia.

Humans become infected by inhaling small droplets generated by showers, fountains, water coolant towers, and air-conditioning systems. It can be said that disease has resulted from developments and improvements in technology, as a result of *L. pneumophila* taking advantage of the new ecological niche we have created. The disease is not spread from person to person.

CLINICAL FEATURES

There are two types of disease caused by *L. pneumophila*. Pontiac fever occurs after a 24- to 48-hour incubation period. It consists of fever, malaise, chills, muscle pains, and headache, but shows very few other signs of respiratory disease. It usually resolves without antibiotic treatment.

The incubation period for Legionnaire's disease is between 2 and 10 days. This is a much more serious infection. It begins in the same way as Pontiac fever, but the patient goes on to develop a cough that can progress to stupor, widespread lung infection, and multi-system organ failure. Because it does not respond to the drugs mainly used in treating pneumonia and is difficult to diagnose, mortality rates are quite high, ranging from between 20 percent and 50 percent. Mortality rates are higher in the elderly or those with existing damage to the lungs or heart.

TREATMENT

Erythromycin, rifampicin, or ciprofloxacin, but usually a combination of two of these.

PREVENTION

No vaccine is available. Prevention is only by ensuring water systems are not overgrown with *L. pneumophila* biofilms. This requires close attention to detail and is not always easy.

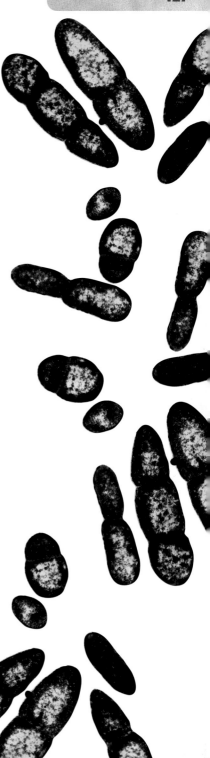

salmonella typhi/typhoid fever

QUICK REFERENCE

■ **ORIGIN**
A solely human pathogenic bacterium.

■ **FIRST IDENTIFIED**
Typhoid was differentiated from typhus in 1782. The bacterium was first isolated in 1885.

■ **TRANSMISSION**
Feco-orally, directly or indirectly in food and water.

■ **PORTAL OF ENTRY**
Gastrointestinal tract.

■ **INCUBATION PERIOD**
10–14 days.

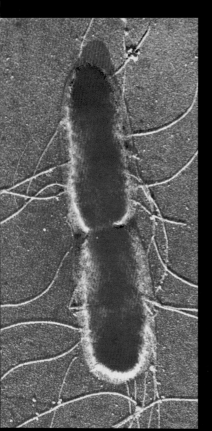

Typhoid fever was given its name because the disease was initially easily confused with typhus. Typhoid, however, is a much more major infection. Although the incidence is falling, it is estimated that there are some 16 million cases worldwide each year, resulting in 600,000 deaths. In some parts of the world, particularly Southeast Asia, the annual incidence is around 1,000 cases per 100,000 of the population. This has been made worse by the bacterium's increasing resistance to antibiotics.

CARRIERS

Salmonella typhi is solely a human path-ogen maintained both by a number of human carriers and by its ability to survive in water. It is estimated that between one in 20,000 and one in 100,000 of the population in countries where typhoid is endemic are carriers.

One notorious carrier was "Typhoid Mary," who in the 1900s worked as a cook for wealthy families in New York, where she caused a number of epidemics. The uncooperative Mary was put in prison, but after a public outcry was released after three years on condition that she would not work again as a cook. She then disappeared, but was found five years later working as a cook in a hospital and still passing on *S. typhi*. Overall, Mary caused 47 cases of typhoid and at least three deaths.

TRANSMISSION

Carriers will have had typhoid at some stage, but do not eliminate the *S. typhi* bacteria. Most often, the bacteria grow in the carrier's gall-bladder and are excreted sporadically throughout his or her lifetime. The infectious dose (the minimum number of bacteria that are needed to cause infection) of *S. typhi* is very low, and it is therefore easily transmitted.

Once ingested, the bacteria pass through the stomach to the small intestine. They then pass through the cells lining the small intestine and are taken to the local lymph nodes where macrophages usually engulf and destroy any bacteria taking this route. However, *S. typhi* bacteria are able to resist and can grow inside the macrophages. They then destroy the macrophages, and escape via the lymphatics to the bloodstream and then around the body. Eventually the bacteria can return to the lymphatic tissue in the small intestine (called Peyer's patches), which can result in potentially fatal intestinal hemorrhage or perforation.

CLINICAL FEATURES
Typhoid has an insidious onset of malaise, chills, headaches, muscle and joint pain, and constipation. This lasts for about a week, during which the patient's temperature rises to between 96.5°F (35.8°C) and 104°F (40°C). By the end of the week the patient is tired, toxic, and confused, becoming increasingly ill. In those who do not recover there is increasing apathy and loss of contact with reality, and in the third week the patient enters the typhoid state. The patient is dull, disoriented, stuporous with toxemia, and close to death.

TREATMENT
S. typhi has become resistant to most of the drugs previously used to treat the various strains and we are fast approaching a time when we can no longer treat typhoid fever in some regions.

PREVENTION
There are two vaccines: a live-attenuated *S. typhi* and another that gives short-term protection. A protein conjugate Vi-antigen vaccine is currently being tested.

salmonella typhimurium/
salmonellosis: food poisoning

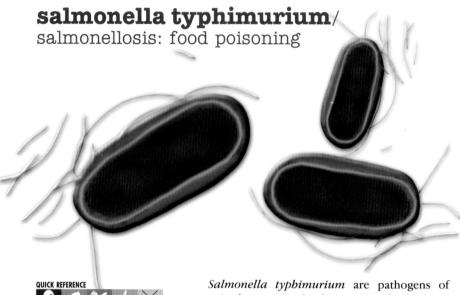

■ ORIGIN
A pathogen of animals transmitted to humans.

■ FIRST IDENTIFIED
The various food-poisoning salmonellae were distinguished from each other in 1929. The upsurge in cases has coincided with industrialized farming.

■ TRANSMISSION
Feco-orally as a food poisoning, or person-to-person.

■ PORTAL OF ENTRY
Gastrointestinal tract.

■ INCUBATION PERIOD
1–5 days.

Salmonella typhimurium are pathogens of animals transmitted to humans as a food poisoning. They infect domestic and wild animals, some of which, especially those kept in large numbers and in close confinement, do not eliminate the bacterium and become persistently infected.

Humans usually become infected by eating fecally contaminated meat, but *S. enteritidis* has also infected the egg-laying apparatus of chickens and it is estimated that approximately one in 700 eggs consumed in the U.K. contains *S. enteritidis* bacteria, though this is not a real problem if the eggs are cooked prior to consumption. Person-to-person spread can also occur.

CLINICAL FEATURES
Illness begins with nausea, anorexia, abdominal pain, and acute watery diarrhea, for up to a week. A small number of patients develop meningitis, septicemia, or even a typhoid-like illness.

TREATMENT
Diarrhea by itself does not need treatment.

PREVENTION
No human vaccine is available. However, vaccines are being used in food animals and the incidence of human salmonellosis is falling in the U.K. and other parts of Europe.

shigella dysenteriae/
bacillary dysentery

All four *Shigella* species—*Shigella dysenteriae, Sh. sonnei, Sh. flexneri,* and *Sh. boydii*—are solely human pathogens. *Sh. dysenteriae* produces the most severe infections and is found predominantly in tropical countries. It produces a toxin that inhibits protein synthesis and can precipitate hemolytic uremic syndrome (HUS—see *Escherichia coli,* overleaf). The others are found in temperate and tropical countries alike. *Sh. sonnei* is the most frequently detected pathogen and is responsible for outbreaks of infection, especially where hygiene is poor.

During the American Civil War there were 1,739,135 cases of bacillary dysentery due to *Sh. dysenteriae* among the combatants with 44,558 deaths. In World War I, nearly 4 percent of casualties in Flanders, and up to 45 percent in East Africa, were due to bacillary dysentery. The infective dose is low, so person-to-person spread occurs easily. In addition, it can be spread on food, in water, and even by flies.

CLINICAL FEATURES

Diarrhea (up to 12 stools per day) is the major feature. The stools are low in volume, and loose rather than watery, and may be accompanied by mucus or blood. In some patients, there may be fever, abdominal pain, and tenesmus (the feeling of still needing to pass a stool directly after emptying due to inflammation of the rectum). Dysentery can be fatal in developing countries due to either dehydration or, less commonly, HUS.

TREATMENT

Treatment is usually required. Antibiotics such as nalidixic acid are used to contain epidemic spread in the tropics.

PREVENTION

No vaccine is available. Improvements in sanitary conditions and personal hygiene help reduce the risk of infection.

QUICK REFERENCE

■ **ORIGIN**
Solely a pathogen of humans.

■ **FIRST IDENTIFIED**
The disease was distinguished from amebic dysentery in 1875. *Sh. dysenteriae* was first isolated in 1898.

■ **TRANSMISSION**
Feco-orally either directly on hands or indirectly in food or water.

■ **PORTAL OF ENTRY**
Intestinal tract.

■ **INCUBATION PERIOD**
Usually 2–3 days.

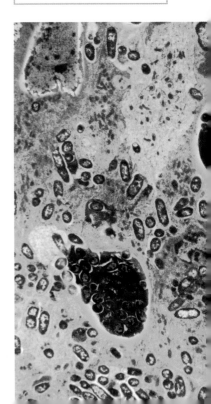

escherichia coli 0157/hemorrhagic
colitis, hemolytic uremic syndrome

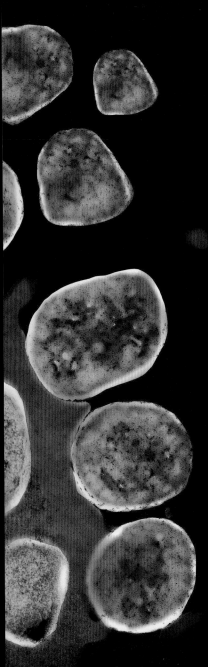

Escherichia coli 0157 is a newly emerged pathogen that is not present in any culture collection of *E. coli* from before 1980. It first emerged in the U.S.A. and Canada where it caused epidemics of hemorrhagic colitis (bloody large-bowel diarrhea) associated with eating hamburgers. Unfortunately, some of the patients infected also developed a life-threatening condition called hemolytic uremic syndrome (HUS), in which the patient's red blood cells become fragmented and the kidneys stop working. The toxins produced by *E. coli* 0157 are among the most powerful known, inhibiting protein synthesis and thus causing the rapid destruction of humans cells.

Along with similar enterohemorrhagic *E. coli* (EHEC) such as *E. coli* 0111 and *E. coli* 026, *E. coli* 0157 bacteria can be found as normal inhabitants of the intestine of cattle, sheep, goats, cats, and dogs, in whom they cause no damage. It is only when humans ingest food or drink contaminated by these host animals that disease occurs.

TRANSMISSION
The first outbreaks in North America and Europe were associated with eating under-cooked hamburgers. When a carrier cow is slaughtered (or even before), its carcass, as well as the carcasses of nearby animals, becomes contaminated with fecal contents containing *E. coli* 0157. This does not matter if the meat is to be grilled, fried, or roasted since the process will kill the *E. coli* 0157 that are on the surface of the meat. However, if the meat is ground, the *E.coli* 0157 are mixed throughout the meat. If the ground meat used to make hamburgers is undercooked (that is, pink) in the center, any *E. coli* 0157 here will have survived and, even in small numbers, is sufficient to initiate disease.

Nowadays, transmission of *E. coli* via hamburgers is rare. However, there are a number of other routes still available to the bacterium, for example storing uncooked meat close to cooked meat (which of course will not be cooked again), drinking unpasteurized milk or contaminated water, or even by not washing hands after patting fecally contaminated cattle and then eating sandwiches.

CLINICAL FEATURES
Patients present with diarrhea, which can vary from mild watery diarrhea to acute diarrhea with bloody stools. Two to three days later, a proportion of patients, but mainly young children and the elderly, may suffer renal shutdown with no urine being passed at all—a manifestation of HUS.

TREATMENT
This is supportive. Some antibiotics potentiate the release of toxin, making the disease worse, and should be avoided.

PREVENTION
No vaccine is available. Prevention is by ensuring the food is free of EHECs, and maintaining good hygiene in butcher's shops and the home.

QUICK REFERENCE

■ **ORIGIN**
A newly evolved bacterium.

■ **FIRST IDENTIFIED**
The first outbreaks were described in 1982 in the U.S.A. and Canada.

■ **TRANSMISSION**
As a food poisoning, or person-to-person feco-orally.

■ **PORTAL OF ENTRY**
Intestinal tract.

■ **INCUBATION PERIOD**
Hemorrhagic colitis: 3–5 days.
HUS: a further 2–3 days.

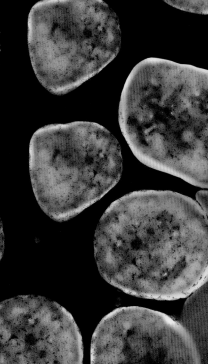

escherichia coli/meningitis, septicemia, urinary tract infection, diarrheal disease

■ **ORIGIN**
The bacteria appear to have evolved with humans.

■ **FIRST IDENTIFIED**
Theodor Escherich first isolated *E. coli* in 1885. The various pathogenic mechanisms have been elucidated over the last 30 years.

■ **TRANSMISSION**
Urinary tract infection: from the intestinal flora. Neonatal infection: from the mother's intestinal and perineal flora. Diarrheagenic *E. coli*: feco-orally, directly or indirectly, in food or water.

■ **PORTALS OF ENTRY**
Intestine or urethra.

■ **INCUBATION PERIOD**
Highly variable.

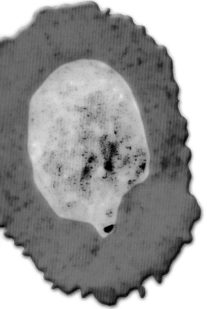

Escherichia coli is part of the normal intestinal flora of many animals including humans, and is usually a harmless commensal. However, some strains of *E. coli* are more pathogenic, causing meningitis and septicemia predominantly in newborn and, especially, premature babies, urinary tract infections, or diarrheal disease.

A number of different *E. coli* strains cause diarrhea. Enterotoxigenic *E. coli* (ETEC) adhere to the small intestinal mucosa and release toxins that cause a voluminous secretory diarrhea. They are an major cause of diarrheal disease in developing countries and in travelers to them.

Enteroinvasive *E. coli* (EIEC) are a minor cause of mild diarrhea. They attach to, invade, and destroy colonic mucosal cells, producing an inflammatory diarrhea. Enteropathogenic *E. coli* (EPEC) are nowadays significant only in developing countries, where they cause acute watery diarrhea that can last for months. Enterohemorrhagic *E. coli* (EHEC) include *E. coli* 0157 (see page 132). Enteroaggregative *E. coli* (EAggEC) are an important cause of pediatric diarrhea. *E. coli* can also cause wound infections.

CLINICAL FEATURES
ETEC, EPEC, and EAggEC produce watery dehydrating diarrhea, and EIEC and EHEC produce bloody diarrhea. Urinary tract infections are characterized by increased frequency of passage of urine, blood in the urine, and a burning pain on passing urine. Loin or back pain and surging high temperatures are a sign that the infection has ascended to the kidneys.

TREATMENT
It is not possible to predict which antimicrobials will work, so treatment is guided by laboratory testing.

PREVENTION
No vaccines are available.

campylobacter spp/
diarrheal disease, food poisoning

Campylobacter jejuni and *C. coli* are part of the normal intestinal flora of a large number of animals, poultry, and other birds, in whom they live harmlessly in the gastrointestinal tract. However, in humans, either can cause diarrheal disease.

Infection is usually through food poisoning. For example, when preparing food, the bacteria on the surface of a raw chicken can be transferred on to hands, cooking implements, and cleaning materials. They are then transferred by hands to mouth, directly, or indirectly by eating uncooked, contaminated food. Infection can also be transmitted in milk and water, by contact with infected domestic pets, and even where birds pecking the tops of milk bottles have defecated, depositing campylobacters into the milk.

CLINICAL FEATURES
Patients develop fever, abdominal pain, and (possibly bloody) diarrhea. A rare complication of Campylobacter enteritis is Guillain-Barré syndrome, which causes damage to motor and sensory peripheral nerves, leading to paralysis and loss of sensation.

TREATMENT
Treatment is not usually required, but antibiotics include erythromycin and ciprofloxacin.

PREVENTION
There is no vaccine and it is unlikely that food animals will ever become campylobacter-free. Good hygiene when handling and preparing food is the best preventative measure.

QUICK REFERENCE

- **ORIGIN**
A harmless commensal of many animal species.

- **FIRST IDENTIFIED**
Campylobacter was first isolated in 1906 as a cause of abortion in sheep. Those causing diarrhea in humans were not isolated until 1975.

- **TRANSMISSION**
Feco-orally directly, but most often as a food poisoning.

- **PORTAL OF ENTRY**
Mouth.

- **INCUBATION PERIOD**
1–7 days.

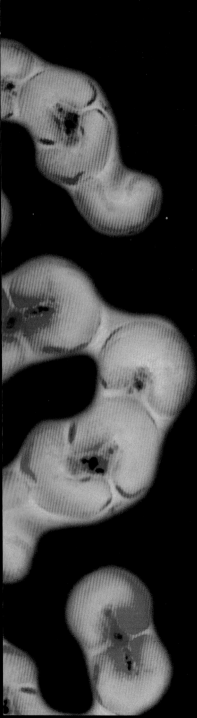

vibrio cholerae/cholera

A cholera-like disease is described in early Chinese, Indian, and Greek literature, but it is probable that the disease did not spread beyond the Indian subcontinent before the beginning of the 19th century.

The strain of *Vibrio cholerae* that causes cholera is *V. cholerae* 01. During a visit to Egypt in 1883, Robert Koch was the first to show that cholera was a bacterial disease. Although it has a relatively high infective dose (over a million bacteria need to be ingested), it causes explosive epidemics of severe dehydrating, acute watery diarrhea, and is endemic in many developing countries.

EPIDEMICS

Since 1817, there have been seven pandemics when *V. cholerae* has emerged from its natural homeland in the Ganges plain and delta and spread around the world. The seventh pandemic began in 1961 in Sulawesi, Indonesia, and spread throughout Southeast Asia. Between 1963 and 1969 it spread to the Indian subcontinent where it displaced preexisting strains. The bacterium entered Africa in the east via the Arabian peninsula, through Djibouti, and to West Africa via the return of infected travelers. By 1978, most countries in central and southern Africa had experienced epidemics. The final stage was in 1991 when it entered South America.

In recent times, the largest epidemics have been in Africa, usually associated with mass camps for refugees displaced by war. In 1994, 42 percent of deaths due to cholera were in Africa, largely as a result of infection in refugees displaced to the Democratic Republic of Congo from the genocide in Rwanda. In one six-week period in 1994, there were 70,000 cases of cholera with 12,000 deaths.

TRANSMISSION

V. cholerae is solely a human pathogen but it also has the ability to survive in watery environments, perhaps in association with aquatic flora and fauna such as plankton. During epidemics, potable water is increasingly contaminated with *V. cholerae*, which means large numbers of people become infected.

The bacterium is ingested in food or water and adheres to the mucosal lining of the small intestine. It then releases at least three toxins: cholera toxin (CT), accessory cholera enterotoxin (ACE), and zonula occludens toxin (ZOT). Of these, CT is the most potent. It adheres to, and penetrates, the enteric mucosal cells and causes them to secrete sodium, chloride, and bicarbonate ions together with water, resulting in the large-volume watery diarrhea.

CLINICAL FEATURES
In its most severe form, cholera presents with a sudden onset of effortless vomiting and voluminous watery diarrhea, usually without abdominal pain. The watery stool is classically described as a rice-water stool. Patients become severely dehydrated and die of shock within 4 to 12 hours. However, there are also between 5 and 40 cases of asymptomatic infection for every infected patient, which can make control of epidemics difficult.

TREATMENT
The mainstay of treatment is to rehydrate the patients orally or intravenously with fluid, electrolytes, and glucose. Antibiotics such as tetracycline or cotrimoxazole decrease the period of diarrhea, but emerging resistance is limiting their value.

PREVENTION
Improved sanitation, mainly separating drinking water and food from human excreta, has eliminated cholera in developed countries. No vaccine is available, but a number are currently being developed.

QUICK REFERENCE

■ **ORIGIN**
A human bacterium that is continuously evolving.

■ **FIRST IDENTIFIED**
V. cholerae 01 was first isolated by Koch in 1883.

■ **TRANSMISSION**
Feco-orally, either directly or indirectly, in food or drink.

■ **PORTALS OF ENTRY**
Mouth and gastrointestinal tract.

■ **INCUBATION PERIOD**
2–3 days.

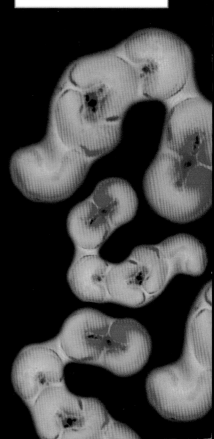

helicobacter pylori/
gastritis, peptic ulcer, stomach cancer

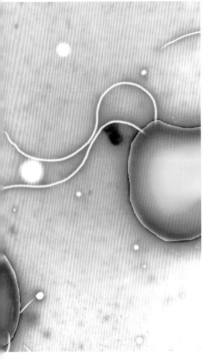

It is estimated that *Helicobacter pylori* colonizes or infects half the world's population. It is solely a human pathogen that has adapted to living in the harsh acidic environment of the human stomach.

Infection rates are related to poor social conditions. In industrialized countries, around half of those aged 60 are infected, but this figure is dropping. In developing countries, up to 80 percent are infected by the age of 10. It is not entirely clear how infection is spread, but *H. pylori* can be detected in saliva, dental plaque, and feces, indicating that both oro-oral and feco-oral spread is possible.

Once *H. pylori* reaches the stomach it burrows beneath the mucin layer and attaches to the cells lining the stomach. Here it is protected from the acid and proteolytic enzymes naturally present in the stomach. This then initiates an inflammatory response (gastritis), which eventually diminishes, but the bacterium persists. After several years, certain strains of *H. pylori* (which possess the genes for virulence) can cause peptic ulcers and even some cancers.

CLINICAL FEATURES

Patients develop abdominal pain, nausea, flatulence, and halitosis (bad breath), which can last up to two months. Months or even years later, peptic ulcer disease may develop. *H. pylori* causes two forms of cancer: carcinoma of the stomach and a tumor of lymphoid tissue in the intestine called maltoma.

TREATMENT

This is usually by a combination of bismuth salts, two or three antibiotics (such as metronidazole, clarithromycin, or ampicillin) and a proton pump inhibitor.

PREVENTION

No vaccine is available.

QUICK REFERENCE

■ **ORIGIN**
A solely human pathogen.

■ **FIRST IDENTIFIED**
First seen by pathologists in the early 1900s but not cultured until 1983.

■ **TRANSMISSION**
Oro-oral or feco-oral.

■ **PORTAL OF ENTRY**
Mouth.

■ **INCUBATION PERIOD**
Acute gastritis: 2–5 days.
Peptic ulceration: months to years.
Gastric carcinoma: decades.

yersinia enterocolitica/
yersiniosis, diarrheal disease

The major host reservoir of the strains of *Yersinia enterocolitica* that cause diarrheal disease is the pig, which carries the bacterium asymptomatically in its gastrointestinal tract. Humans usually become infected by contact with pigs, eating improperly cooked pork, or poor kitchen hygiene.

Y. enterocolitica is rarely found in hot climates but its prevalence increases in colder countries, for example, Scandinavia. Thus it also tends to be more prevalent in the winter months. It can also be found, though less commonly, in sheep, goats, cattle, and the environment. Infection via water and milk and person-to-person transmission are also possible. Most infections are in children under the age of five.

CLINICAL FEATURES
Yersiniosis causes mild to severe enteritis with diarrhea, which can be bloody. It may also cause terminal ileitis, the appearance of which is similar to acute appendicitis. The infection can be short-lived, can relapse, or persist for weeks.

TREATMENT
Treatment is not needed for mild uncomplicated disease, but bacterium is sensitive to gentamicin, cotrimoxazole, tetracycline, and ciprofloxacin.

PREVENTION
No vaccine is available. Good hygiene is the mainstay of prevention.

QUICK REFERENCE

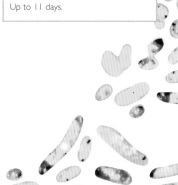

- **ORIGIN**
  A harmless commensal of pigs.

- **FIRST IDENTIFIED**
  Bacterium first described in 1939.

- **TRANSMISSION**
  Feco-orally as a food poisoning, or person-to-person.

- **PORTAL OF ENTRY**
  The mouth to the lower small bowel.

- **INCUBATION PERIOD**
  Up to 11 days.

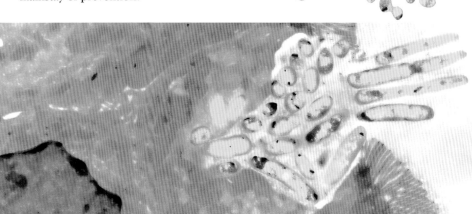

yersinia pestis/plague

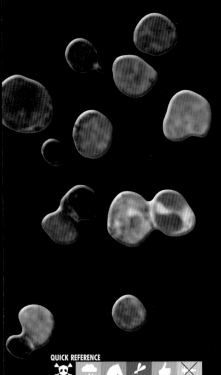

Plague is an ancient disease, but it is still endemic in many parts of the world. It is a zoonotic infection, spread from its rodent hosts to humans by fleas, although person-to-person spread is also possible. Although the brown rat and the black rat are important for transmission to humans, plague kills them and they are probably not the definitive host. *Y. pestis* has been isolated from more than 230 other rodent species. Even pets such as dogs and cats can be infected, usually by eating an infected rodent.

PANDEMICS

There have been three pandemics of plague. The first (Justinian) plague began in 542 A.D. It probably escaped from its homeland in the Himalayan borders of India and China, perhaps via trade routes spreading to the Eastern Roman Empire in Constantinople. There, it is estimated to have killed over 10,000 people in one day. It spread westward through Italy and southern France, and out to Britain by 544 A.D. Overall, it is estimated to have killed 100 million people and ushered in the Dark Ages. It is thought to have vanished from Europe in 767 A.D., not to return for another 600 years.

The next pandemic was the Black Death, which began in 1348, having spread across the Asian steppes along southern Muslim trade routes. It was also an early example of biowarfare. The Tartar army, besieging the city of Kaffa (in present-day Ukraine), catapulted corpses of plague victims into the city to create panic and infection. The Black Death returned in a series of epidemics, the last of which was in Marseille in 1720.

The third and current pandemic began in Yunan Province, China, in 1860 and reached Hong Kong in 1894. It was then spread around the world by black rats carried on ships.

Plague is still reported in a number of countries in Africa, Asia, and South America, and in the U.S.A. In the 1990s, an average of 2,650 cases were reported to the World Health Organization each year.

QUICK REFERENCE

■ ORIGIN
A bacterium of rodents accidentally transmitted to humans by fleas.

■ FIRST IDENTIFIED
An ancient disease (pestilence). The bacterium was first isolated in 1894.

■ TRANSMISSION
By flea bite from rodent, or by inhalation, person-to-person.

■ PORTALS OF ENTRY
Through the skin from fleas (bubonic plague), or into the lungs by inhalation (pneumonic plague).

■ INCUBATION PERIOD
Usually 2–5 days but can be as long as 15 days.

TRANSMISSION

The flea (particularly the rat flea) is the vector for transmission. The human flea can also transmit plague but is less efficient at doing so. Fleas bite infected rats to obtain a blood meal, but also acquire *Y. pestis*, which prevents these blood meals from being absorbed. The flea then becomes increasingly hungry and bites more frequently. At each bite, *Y. pestis* is transferred to a new host, be this a rat or a human.

CLINICAL FEATURES

Bubonic plague is the main form of the disease. It begins with a sudden onset of fever, rigors, and headache. The bubos, which are enlarged lymph nodes and swelling of the surrounding tissue, appear within 36 hours. The bubos are painful and tender, and the overlying skin is red and inflamed. Initially the bubos are firm or rubbery, but as the disease progresses, they become fluctuant and may burst through the skin.

The patient becomes severely ill, and by the second or third day a purpuric (bruising) skin rash appears. This is purple-black, which may have been the origin of the name, Black Death. Most deaths occur within the third to the sixth day of illness, and if untreated, up to half of those with bubonic plague will die. All patients become septicemic, and the term septicemic plague is reserved for those who get overwhelming illness, often without bubo formation.

In a proportion of patients, *Y. pestis* reaches the lungs via the bloodstream and causes bronchopneumonia. Such patients will cough out large numbers of bacteria and transmit infection via the airborne route, person-to-person. This is pneumonic plague.

TREATMENT

Prompt administration of antimicrobials can decrease mortality rates to below 15 percent. However, antibiotic resistance is increasing.

PREVENTION

No vaccine is available.

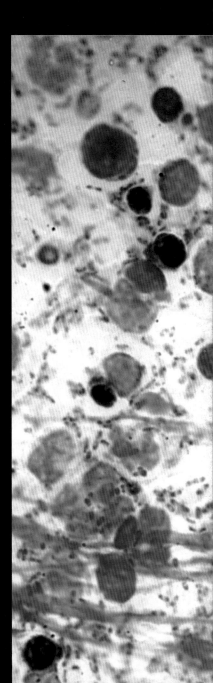

klebsiella pneumoniae/urinary tract infection, pneumonia, septicemia

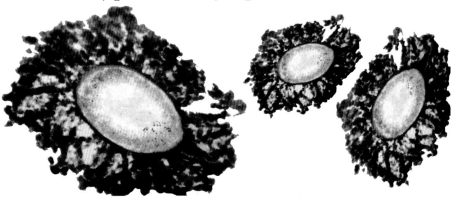

QUICK REFERENCE

■ **ORIGIN**
A bacterium usually found in the intestine of humans and other animals.

■ **FIRST IDENTIFIED**
Bacterium first isolated in 1883.

■ **TRANSMISSION**
Person-to-person feco-orally on hands or food.

■ **PORTAL OF ENTRY**
The mouth to the intestinal tract, thence to infected sites.

■ **INCUBATION PERIOD**
Not applicable.

Klebsiella pneumoniae is named after Erwin Klebs (1834–1913), a German bacteriologist. Originally the bacterium was described in 1883 as a cause of severe pneumonia and called Friedlander's bacillus. This is uncommon now.

K. pneumoniae and its various subspecies are part of the normal human intestinal flora but are present in only relatively small numbers (around 100 bacteria per gram of feces). Most often it is associated with nosocomial (hospital-acquired) infection, and generally occurs when a patient's immune system is not working well.

The infections are usually endogenous (derived from the patient's own microflora) and are often associated with the use of urinary catheters or artificial ventilation. *K. pneumoniae* appears to be particularly adept at acquiring antibiotic-resistance genes, which can pose problems in treating infections.

CLINICAL FEATURES
Dependent on the site of infection.

TREATMENT
Treatment is usually guided by testing antibiotic susceptibility in the laboratory.

PREVENTION
No vaccine is available.

klebsiella granulomatis/
donovanosis, granuloma inguinale

Klebsiella granulomatis has only recently been assigned to the genus *Klebsiella*. It was previously called *Donovania granulomatis* and after that *Calymmatobacterium granulomatis*. The reason for the confusion is that it has only just been cultured artificially and is an obligate intracellular bacterium only growing inside human cells.

Granuloma inguinale is a disease characterized by long-lasting genital ulceration that can result in severe genital destruction. It is sexually transmitted and found predominantly in the tropics and subtropics in India, Papua New Guinea, Northern Australia, Southern China, Africa, Brazil, Argentina, Mexico, and some southern states of the U.S.A.

The transmission rate between sexual partners is less than 50 percent but epidemics of infection have been described in Papua New Guinea following ritual male homosexual puberty rites. Nonsexual transmission (mother-to-baby or by accidental inoculation) is rare.

CLINICAL FEATURES
A small lump or papule develops at the site of inoculation on the genitalia. The bacteria then move via the lymphatics to the draining lymph nodes. The initial papule evolves to become a beefy red ulcer which is usually painless. The bacteria in the lymph nodes burst through the overlying skin to cause further ulceration. Blockage and elephantasis are not uncommon. Genital mutilation such as autoamputation of the penis and carcinoma of the skin may develop as long-term complications.

TREATMENT
If treatment with doxycycline, erythromycin, or cotrimoxazole is started sufficiently early in the disease process, there is a good chance of cure.

PREVENTION
No vaccine is available. Prevention is by safe sex.

QUICK REFERENCE

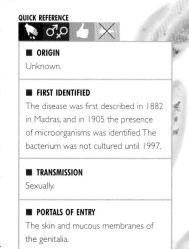

■ **ORIGIN**
Unknown.

■ **FIRST IDENTIFIED**
The disease was first described in 1882 in Madras, and in 1905 the presence of microorganisms was identified. The bacterium was not cultured until 1997.

■ **TRANSMISSION**
Sexually.

■ **PORTALS OF ENTRY**
The skin and mucous membranes of the genitalia.

■ **INCUBATION PERIOD**
3–90 days (most often between 1 and 6 weeks).

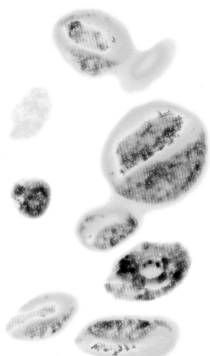

klebsiella rhinoscleromatis/
rhinoscleroma

QUICK REFERENCE

■ **ORIGIN**
Unknown, but the disease has been present for at least 1,500 years.

■ **FIRST IDENTIFIED**
The first accurate clinical description was in 1870. *K. rhinoscleromatis* was first isolated in 1882.

■ **TRANSMISSION**
Unknown.

■ **PORTALS OF ENTRY**
The junction sites between the skin and mucous membranes of the face.

■ **INCUBATION PERIOD**
Unknown.

Klebsiella rhinoscleromatis causes a chronic progressive granulomatous disease of the upper airways. Although described as a rare disease, more than 16,000 cases were reported between 1960 and 1992, and this is an underestimate.

Rhinoscleroma is found in rural areas, especially where socioeconomic conditions are poor. Females are 13 times more likely to develop disease and most often rhinoscleroma presents in the second or third decades of life. It is widely reported from the Middle East, tropical Africa, India, Southeast Asia, and Central and South America. In Europe, it is found in Russia, Poland, Hungary, and Romania. Little is known about how it is spread.

CLINICAL FEATURES
This chronic disfiguring and debilitating infection affects most parts of the upper airways but the nose is affected in almost all cases and the pharynx in up to half of cases. It can spread by extension to the brain. Lesions appear as bluish-red rubbery granulomas. These then enlarge and there may be destruction of the underlying bone. They can persist for years.

TREATMENT
This involves surgery to remove granulomatous tissue, plus antimicrobials such as ciprofloxacin. Patients may be left with severe scarring.

PREVENTION
No vaccine or preventative interventions are currently available.

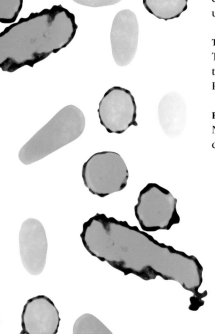
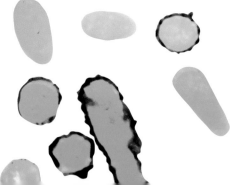

citrobacter, enterobacter, serratia/
hospital-associated infections

These three bacterial genera are part of the enteric flora of humans and other animals. There are 11 different *Citrobacter* species but the most important are *Citrobacter freundii* and *C. koseri*. The latter is associated with meningitis and brain abscesses in premature newborn babies.

There are also 11 enterobacter species, and of these *Enterobacter cloacae* and *Ent. sakazakii* cause human infections. *Ent. sakazakii* has recently been found to cause outbreaks of severe infection in debilitated infants consuming milk-formula feeds.

Serratia marcescens was first described in 1923 as a cause of the red discoloration of polenta (a cornmeal porridge), and has probably been responsible for some of the "miracles" throughout history when "blood" has appeared on foodstuffs. The red pigment released by the bacterium is called prodigiosin and is currently being investigated as a potential anticancer drug. *S. marcescens* is also used as a biological tracer because of its pigment production.

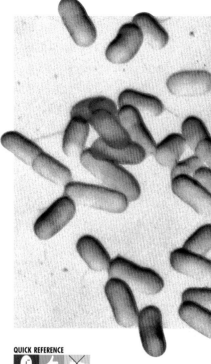

CLINICAL FEATURES
All of the bacteria are predominantly causes of hospital-associated infections. They cause wound infections, pneumonia in ventilated patients, urinary tract infections, septicemia, and meningitis.

TREATMENT
The bacteria are not predictably sensitive to any antimicrobial, therefore treatment is guided by sensitivities determined in the laboratory.

PREVENTION
No vaccines are available.

QUICK REFERENCE

■ **ORIGIN**
Normal intestinal commensals of humans and other animals that are also found in the environment.

■ **FIRST IDENTIFIED**
Citrobacter freundii was first described in 1932, *Enterobacter cloacae* in 1925, and *Serratia marcescens* in 1923.

■ **TRANSMISSION**
Feco-orally directly or on hands or in food.

■ **PORTAL OF ENTRY**
The mouth to the intestinal tract and from there to sites of infection.

■ **INCUBATION PERIOD**
Not applicable.

proteus mirabilis/
urinary tract infection

QUICK REFERENCE

■ **ORIGIN**
A normal intestinal commensal of humans and other animals.

■ **FIRST IDENTIFIED**
First isolated in 1885.

■ **TRANSMISSION**
An endogenous infection.

■ **PORTAL OF ENTRY**
The urethra to the bladder and beyond.

■ **INCUBATION PERIOD**
Not applicable.

Proteus mirabilis is a highly motile bacterium that is part of the intestinal flora of humans and other animals, and can also be found in decomposing meat and sewage. It possesses a powerful enzyme that breaks down urea (present in large quantities in urine) to form ammonia, and is therefore a major cause of urinary tract infections.

The bacteria can colonize the perineum and urethra. If conditions are favorable, *P. mirabilis* then ascends the urethra to reach the bladder. Here it is bathed in nutrients (especially urea). The breakdown of urea to ammonia creates alkaline conditions and this may precipitate the formation of stones in the bladder, ureters, or kidneys.

CLINICAL FEATURES
Urinary tract infections present as lower abdominal pain, pain on urination, and blood in the urine. If the bacteria ascend beyond the bladder the patient will also develop high, swinging temperatures and loin pain.

TREATMENT
It is not possible to predict which antibiotic will work as *P. mirabilis* can be resistant. Laboratory testing usually therefore guides therapy.

PREVENTION
No vaccine is available, nor likely to be.

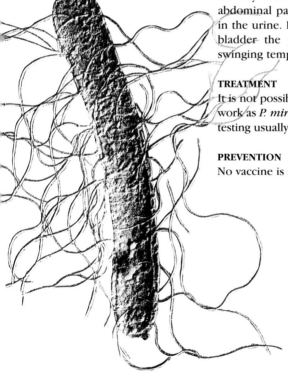

burkholderia cepacia/
lung infection in cystic fibrosis

Burkholderia cepacia is one of a number of species in the recently described genus *Burkholderia*. It was originally called *Pseudomonas cepacia*, but it is not closely related to other *Pseudomonas* species, so the genus was renamed *Burkholderia* in honor of Burkholder, who first isolated the bacterium as a cause of onion-rot in 1950. The species name is derived from the Latin for onion (*cepa*).

There are more than 11 different species in the *B. cepacia* complex, which rarely cause problems in humans. However, of these, *B. cenocepacia* is the major pathogen. It is particularly associated with severe lung infections in patients with the genetic disorder, cystic fibrosis.

B. cenocepacia is able to transmit person-to-person presumably via respiratory secretion, and in up to 40 percent of patients it produces lung damage and septicemia, which is rapidly fatal. However, in some cystic fibrosis patients, it remains in the lung without causing problems, for reasons that are as yet unknown.

CLINICAL FEATURES

Patients with cystic fibrosis produce thick viscid sputum that can block the airways. *B. cenocepacia* infection leads to an exacerbation of the condition with increased production of purulent sputum and a fever. It produces continuing decreased respiratory function or can progress rapidly to septicemia and death.

TREATMENT

B. cepacia bacteria is resistant to a wide range of antibiotics and treatment is difficult.

PREVENTION

No vaccine is available. Spread has been prevented by strict segregation of cystic fibrosis patients infected with *B. cenocepacia* from uninfected ones in hospitals and at social events.

QUICK REFERENCE

■ **ORIGIN**
An environmental bacterium.

■ **FIRST IDENTIFIED**
B. cepacia was first described in 1950. Its association with infection in cystic fibrosis was first described in 1984.

■ **TRANSMISSION**
Person-to-person on hands or airborne.

■ **PORTAL OF ENTRY**
Through the upper airways to the lungs.

■ **INCUBATION PERIOD**
Not applicable.

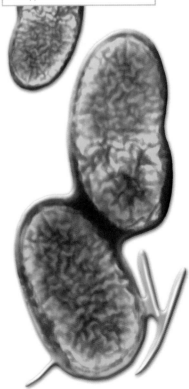

burkholderia pseudomallei/
melioidosis

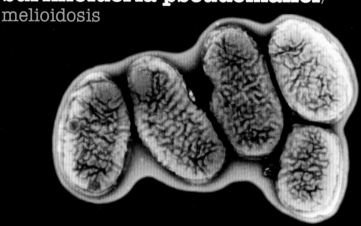

QUICK REFERENCE

■ **ORIGIN**
An environmental bacterium.

■ **FIRST IDENTIFIED**
Melioidosis was first described in Rangoon in 1912.

■ **TRANSMISSION**
From the moist environment through skin abrasions or cuts, or by inhalation.

■ **PORTALS OF ENTRY**
Through the skin or by inhalation through the respiratory tract.

■ **INCUBATION PERIOD**
Up to 21 days (average 6 days).

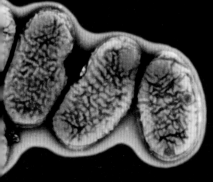

Burkholderia pseudomallei is easily isolated from mud and surface water (especially rice paddy) in endemic areas such as Singapore, Thailand, and northern Australia. Humans probably become infected by inhalation of the bacteria or through cuts when barefoot in mud or water containing the bacteria. These bacteria can persist silently for years in human hosts and then reactivate causing melioidosis. *B. pseudomallei* has recurred in U.S. veterans of the Vietnam war so often that it is known as the Vietnamese time bomb. It is considered a potential biowarfare weapon.

CLINICAL FEATURES
Not all of those who acquire infection develop disease, but factors such as diabetes, chronic renal failure, liver disease, and pregnancy make disease more likely. Disease in healthy hosts presents as a flu-like illness. In those with overt melioidosis, half have septicemia. The remainder have pneumonia, meningitis, or abscesses in major organs. Untreated, severe melioidosis (septicemia) has a high mortality rate.

TREATMENT
A high intravenous dose of ceftazidime for at least two weeks decreases the mortality rate.

PREVENTION
No vaccine is available.

pseudomonas aeruginosa/
nosocomial infections, lung infections in cystic fibrosis

Although ubiquitous in the moist environment, *Pseudomonas aeruginosa* is not usually found in the sea. It generally causes infection only when there is some impairment of immunity.

Only between 2 percent and 10 percent of humans outside hospitals carry *P. aeruginosa* in their intestines, and there is evidence that this is more likely in vegetarians, presumably because they are more likely to consume foods containing *P. aeruginosa*.

In hospitals, up to 60 percent of patients carry *P. aeruginosa*, mostly in the intestine. The longer that patients are in hospital and the more antibiotics they are given, the greater the chance that they will become carriers of the bacteria.

Considering the vast array of virulence factors and toxins produced by *P. aeruginosa*, it is surprising that it is not a primary pathogen. The main infections it causes are those of the skin and eye, wounds and burns, bones and joints, and chest infections in patients with cystic fibrosis (in whom it causes exacerbation and gradual loss of lung function). In the first three, there is usually damage to tissue before infection begins.

CLINICAL FEATURES
These depend upon the site of infection. Folliculitis is the main skin infection and this occurs when subjects immerse themselves in hot tubs or jacuzzis that, if improperly maintained, may contain *P. aeruginosa*. The skin becomes water-logged and *P. aeruginosa* gains access, resulting in a widespread maculopapular or vasiculopustular rash that is unsightly but not life-threatening.

TREATMENT
Guided by antibiotic sensitivity testing.

PREVENTION
No vaccine is available.

QUICK REFERENCE

■ **ORIGIN**
A ubiquitous environmental bacterium.

■ **FIRST IDENTIFIED**
Infection was first described in 1850.
P. aeruginosa was first cultured in 1882.

■ **TRANSMISSION**
From the moist environment.

■ **PORTAL OF ENTRY**
Many (see main text).

■ **INCUBATION PERIOD**
Not applicable.

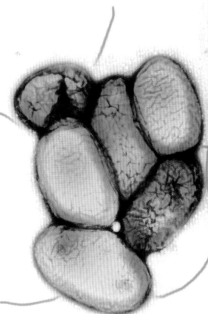

spirillum minus/
rat-bite fever

Spirillum minus has yet to be grown in the laboratory in artificial culture, but can be cultured in mice or rats. It is naturally carried by rats but other animals such as cats, ferrets, and bandicoots have transmitted infection to humans by biting.

Most cases have been in Japan (where it was discovered in 1908 and named Sodoku), but there have also been cases in Australia, North and South America, Africa, and Europe. It is not a major disease because it requires that a human is bitten by a wild rat or other animal colonized by the bacterium. Rarely, humans are infected by consuming food or drink contaminated by rat urine.

CLINICAL FEATURES

Sodoku is a relapsing fever that may subside spontaneously or persist for many months. Most often, but not always, there is a history of a rat bite that heals, but when the disease starts, it breaks down to form an ulcer and the draining lymph nodes become swollen and painful. The fever rises over three days to 104°F (40°C), ends by crisis after three days, to be followed by a quiescent period of between five and 10 days. The process then starts again. With each bout of fever, the site of the initial bite ulcerates further. The mortality rate is 10 percent.

TREATMENT

Penicillin by injection works well.

PREVENTION

No vaccine is available. Prevention is by avoiding contact with rats or their urine.

QUICK REFERENCE

■ **ORIGIN**
A normal bacterium of rats and some other animals.

■ **FIRST IDENTIFIED**
Disease first described in Japan in 1908.

■ **TRANSMISSION**
By rat (or, occasionally, other animal) bite or by rat urine contaminating food or drink.

■ **PORTALS OF ENTRY**
Through the skin by bite, or less commonly through the gastrointestinal tract.

■ **INCUBATION PERIOD**
5–30 days (but usually 5–10 days).

streptobacillus moniliformis/
haverhill fever, rat-bite fever

Streptobacillus moniliformis has had a number of name changes and has recently been renamed *Acpinobacillus muris*. It is a more common cause of rat-bite fever than *Spirillum minus* (see opposite). However, it is not always transmitted by rat bite and a number of outbreaks have been associated with consumption of raw (unpasteurized) milk contaminated by rat urine. This form is called Haverhill fever (named after the town in Massachusetts where an epidemic occurred) as there is no bite ulcer.

Human infection has been described in the U.S.A. and Canada, Brazil, Mexico, Paraguay, Europe (Denmark, Finland, France, Germany, Greece, Holland, Italy, Norway, Spain, Sweden, and the U.K.), Australia, and India. These are mostly sporadic cases but epidemics have also occurred (for example, in the U.K.). Rats are the major reservoir and carry the bacteria asymptomatically in the nasopharynx, but other animals such as mice, gerbils, squirrels, and koala bears can also be carriers.

CLINICAL FEATURES
The clinical features of infection acquired by bite and ingestion are similar, except that the ingestion-acquired form presents a sore throat. Unlike Sodoku, this form of rat-bite fever is not accompanied by ulceration of the bite wound.

The patient develops high fever, prostration, headache, and a generalized rash resembling measles, with enlarged lymph nodes. In untreated cases it can subside spontaneously, but in most, it is relapsing with night sweats and high fever coming and going for months. The mortality rate is around 10 percent.

TREATMENT
The antibiotic tetracycline.

PREVENTION
No vaccine is available.

QUICK REFERENCE

■ **ORIGIN**
A bacterium carried by rats and some other animals.

■ **FIRST DESCRIBED**
Disease identified in 1839, bacterium first isolated in 1926.

■ **TRANSMISSION**
By rat bite or consumption of rat urine.

■ **PORTAL OF ENTRY**
Through the skin by bite or through the intestine.

■ **INCUBATION**
3-10 days.

francisella tularensis/tularemia

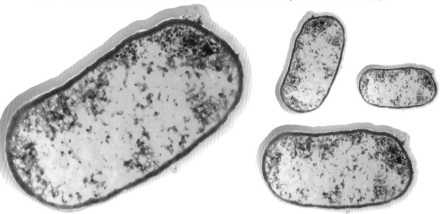

■ **ORIGIN**
A bacterium present in a number of wild animals.

■ **FIRST IDENTIFIED**
The disease was first described in Japan in 1837. *F. tularensis* was first isolated in 1911 in the U.S.A.

■ **TRANSMISSION**
By insect bite (ticks, deerfly, or fleas), or by animal bite, inoculation into cuts, ingestion, or inhalation.

■ **PORTALS OF ENTRY**
The skin, conjunctivae, lungs, or gastrointestinal tract.

■ **INCUBATION PERIOD**
Up to 10 days.

Tularemia, a zoonosis, is caused by *Francisella tularensis*. Person-to-person transmission does not occur. Humans become infected directly from asymptomatic host reservoir animals during butchering; by ingestion of the poorly cooked meat of a reservoir host; or from the bite of an infected carnivore. They can also become infected from contaminated water, dust, or hay, by ingestion or inhalation. The main insect vectors are ticks and tabanid (blood-sucking) flies.

CLINICAL FEATURES
Infection is often asymptomatic. The most common, cutaneous form is caused by insect or animal bite; a pustule develops, leaving a punched-out ulcer. There is often painful enlargement of the draining lymph nodes, for up to three months.

Ingestion of *F. tularensis* leads to the oral and abdominal forms of the disease with necrotizing pharyngitis or abscesses on the palate. If the bacteria lodge lower down in the intestines they can cause peritonitis. *F. tularensis* can spread, causing septicemia or a typhoid-like illness, which can lead to fatal pneumonia or meningitis.

TREATMENT
Intramuscular streptomycin or gentamicin.

PREVENTION
No vaccine is available.

aeromonas hydrophila/
wound infection, diarrheal disease

Together with *Aeromonas sobria* and *A. caviae,* *A. hydrophila* forms a group called the aeromonads that are ubiquitous in the moist environment, including streams, rivers, and lakes. Their numbers are highest at the end of summer (around 105 per milliliter of water). They can be found in the feces of between 2 percent and 4 percent of apparently healthy people.

A. *hydrophila* possesses an excess of virulence determinants but does not appear to be particularly pathogenic to humans. Aeromonads have been isolated more frequently from the stools of patients with diarrhea than from controls, but this does not necessarily prove that they cause diarrhea.

If aeromonads gain access (often through infected bathing water) to normally sterile organs around the body, they can cause skin and soft-tissue infections, eye infections, peritonitis, or liver abscesses. In patients with a solid tumor or leukemia, they can cause of septicemia and meningitis.

CLINICAL FEATURES
No specific features for *A. hydrophila*.

TREATMENT
Aeromonads are usually sensitive to gentamicin, ceftazidime, cotrimoxazole, and ciprofloxacin. However, antibiotic resistance is increasing.

PREVENTION
No vaccine is available. Avoiding contact with the bacteria is the only preventative measure.

QUICK REFERENCE

- **ORIGIN**
An environmental bacterium.

- **FIRST IDENTIFIED**
The *Aeromonas* genus was first described in 1936.

- **TRANSMISSION**
Through contact with water or foods contaminated with the bacteria.

- **PORTALS OF ENTRY**
Usually through the skin but can be by ingestion.

- **INCUBATION PERIOD**
Not applicable.

plesiomonas shigelloides/
diarrheal disease

Plesiomonas shigelloides is related to the aeromonads. It was first described as *Plesiomonas* in 1962, but had been known since 1947. It has been found in the intestines of fish in Japan and Europe but is also found in large numbers in surface waters in warm months. Once the temperature drops below 45°F (7°C) it is no longer detectable. *P. shigelloides* is rarely detected, if ever, in apparently healthy humans. Little is known of any pathogen-causing determinants.

CLINICAL FEATURES
Like the aeromonads, there is some dispute over the role of *P. shigelloides* in diarrheal disease. It has been linked to a number of outbreaks of diarrhea in Africa, India, and Japan where the same *P. shigelloides* has been isolated from each of the patients and their drinking water. It is definitely not a major cause of diarrheal disease.

TREATMENT
Treatment is not necessary.

PREVENTION
No vaccine is available, nor likely.

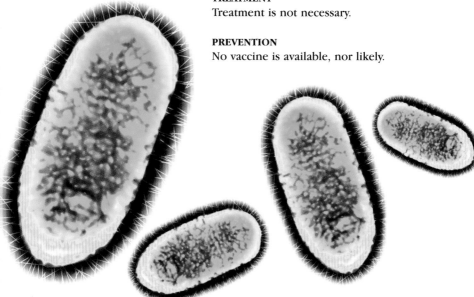

borrelia burgdorferi/lyme disease

The initial presentation of Lyme disease is erythema migrans (EM). In 1977, a cluster of cases of EM and chronic arthritis in Old Lyme in Connecticut, U.S.A, became known as Lyme arthritis. However, it soon became apparent that EM was also associated with heart and brain damage, thus the name was changed to Lyme disease.

Small mammals such as mice, voles, squirrels, and racoons are the definitive hosts and have bacteremia but apparently no disease. The bacteria are transmitted between animals and from animals to humans by hard ixodid ticks. Lyme disease has been detected in North America, Europe, Asia, China, and Japan.

CLINICAL FEATURES
Between three days and four months after an infected tick bite, around 75 percent of patients present with EM. The remainder have no recollection of a bite (usually because the bite is from a tiny nymph) and present only with Lyme disease.

EM is characterized by an expanding ring of red inflamed rash. There may be fever, headaches, chills, malaise, and enlarged lymph nodes, which may persist for up to a year. At some point, there are usually joint pains and swelling, sometimes resulting in chronic arthritis.

In most individuals, *B. burgdorferi* also invades the central nervous system causing diseases from acute meningitis to chronic encephalopathy with loss of short-term memory. In up to 10 percent of cases, there is myocarditis, pericarditis, or conduction disorders of nerve impulses from the heart atrium to the ventricles.

TREATMENT
There is no proven, universal treatment regimen. Penicillins, cephalosporins, and tetracycline kill *B. burgdorferi* in vitro and appear to work in treating EM. Treatment is usually given for 14 to 28 days, most often using ceftriaxone.

PREVENTION
No vaccine is available. Prevention is by avoiding contact with ticks.

QUICK REFERENCE

■ **ORIGIN**
A bacterium found in small mammals.

■ **FIRST IDENTIFIED**
Erythema migrans was first recognized in 1909, but *B. burgdorferi* was not cultured until 1981.

■ **TRANSMISSION**
By ixodid ticks.

■ **PORTAL OF ENTRY**
Through the skin.

■ **INCUBATION PERIOD**
Erythema migrans: 3 days to 4 months. Other manifestations occur weeks to months later.

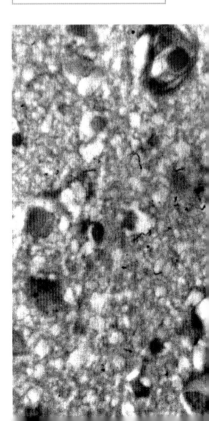

borrelia recurrentis/relapsing fever

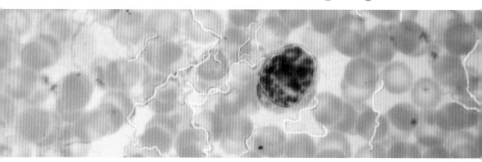

QUICK REFERENCE

■ **ORIGIN**

A human bacterium of unknown origin.

■ **FIRST IDENTIFIED**

The diseases and their vectors were described in 1891 and 1905 for louse-borne and tick-borne relapsing fever respectively. These bacteria were not cultured artificially until 1971.

■ **TRANSMISSION**

By insect vectors: lice and ticks.

■ **PORTAL OF ENTRY**

Through the skin by blood-feeding insects.

■ **INCUBATION PERIOD**

Tick-borne relapsing fever: usually 1–2 days but up to 14 days. Louse-borne relapsing fever: 4–8 days.

Borrelia recurrentis causes louse-borne relapsing fever and *B. duttoni* causes East African tick-borne relapsing fever. It is estimated that there were 15 million cases of louse-borne relapsing fever in sub-Saharan Africa, Sudan, Ethiopia, Eastern Europe, and Russia from 1910 to 1945, resulting in more than five million deaths.

It is a disease of war and poverty, which facilitate heavy louse infestation. The infection cycles between humans and insects and back again are caused by insect bites. The bacteria then enter the bloodstream, causing a recurring fever.

CLINICAL FEATURES

Tick-borne relapsing fever is generally milder than the louse-borne fever, which begins suddenly with chills and a rapid rise in temperature to 106°F (41°C) or above, when the patient becomes apathetic and quite ill. After five to seven days the temperature falls by crisis and there may be a state of collapse.

Relapse occurs in two-thirds of patients. A second relapse occurs in a quarter of cases. More than four relapses is rare. Death usually occurs in the first episode of fever in between 2 per cent and 9 percent of patients.

TREATMENT

Intravenous tetracycline given alone, or together with penicillin, is effective, but killing the *Borrelia* can result in a severe reaction.

PREVENTION

No vaccine is available.

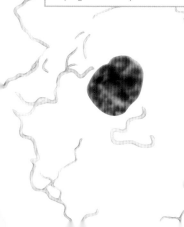

leptospira interrogans/
leptospirosis, weil's disease

There are around 200 different variants of *Leptospira interrogans*, found in animals including dogs and cattle. All are zoonoses, and onward person-to-person spread is rare (though bacteria can cross the placenta to infect the fetus). In some cases, the reservoir host develops disease.

The bacteria are excreted in the host's urine, and can survive for long periods in water. Humans become infected by bacteria in streams and ponds, for example, penetrating through cuts or grazes or even intact mucous membrane. Leptospirosis is more prevalent in developing countries, but is often asymptomatic. The severe form of leptospirosis is Weil's disease.

CLINICAL FEATURES
Disease begins as a "flu-like" illness with fever, chills, sore throat, headaches, muscle and back pain, and nausea. In severe cases, the temperature remains high, and the liver enlarges and begins to fail, causing jaundice. There is bleeding into the skin and lungs. The cause of death is usually kidney failure, and the mortality rate varies from 2 percent to 11 percent.

TREATMENT
Intramuscular penicillin.

PREVENTION
No vaccine is available.

QUICK REFERENCE

■ **ORIGIN**
A bacterium principally of rodents but one that can be transmitted to humans and other animals.

■ **FIRST IDENTIFIED**
Weil's disease was first described in 1886, *L. interrogans* between 1914 and 1916.

■ **TRANSMISSION**
Through the urine of infected animals in surface water.

■ **PORTALS OF ENTRY**
Through cuts and abrasions in skin or through mucous membranes.

■ **INCUBATION PERIOD**
7–12 days.

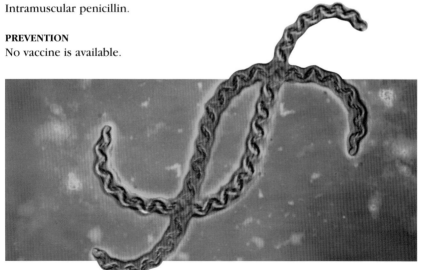

treponema pallidum/syphilis

Treponema pallidum is solely a human pathogen. It has never been grown in artificial culture, but will grow if inoculated into rabbit skin. It takes as few as 50 to 100 bacteria to initiate infection in a rabbit.

Syphilis was first observed in Europe at the end of the 15th century. Between 1493 and 1530 Europe suffered a large epidemic of this new disease which at first was called *morbus gallicus*, or the "French pox," since it appeared to be spread by a French army invading Italy. In 1530, Fracastoro coined the name syphilis in his poem "Syphilis, sive morbus gallicus," but the name did not become generally accepted until the 18th century.

It was accepted that syphilis was a sexually transmitted, or venereal, disease in the 1520s. However, *T. pallidum* is one of a limited number of pathogens that can cross the placenta to infect the fetus, leading to congenital syphilis. This occurs only if the mother has primary or secondary syphilis, or early latent syphilis.

ORIGIN
The origin of the disease is a subject of intense debate. The fact that the first description of syphilis and the onset of a large epidemic coincided with the return of Christopher Columbus from the New World in 1493 has led to suggestions that it was a disease endemic in the Caribbean. It has been claimed that on arrival in the Caribbean, the crew, having spent a long time at sea, became infected during sexual contact with locals, and then brought it back to Europe. Others claim that a mild endemic form of the disease somehow became much more virulent.

Syphilis was greatly feared because it progresses inexorably, producing neurological and psychiatric manifestations and eventually death. It wasn't until World War II that an effective cure, penicillin, became available. The disease is still a scourge, especially in developing countries. In developed countries, it declined throughout the 20th century with only brief rises at the end of both World Wars. However, it is showing something of a resurgence in, for example, the former Soviet Union, and with increasing ease of international travel, cases are on the increase in Europe and the U.S.A.

CLINICAL FEATURES

The primary lesion is called a primary chancre and appears at the site of inoculation of the *T. pallidum*. It consists of a firm ulcer with a raised edge, which is usually painless, does not bleed, and has a clean base. Patients are infectious at this time. The chancre can be anywhere on the genitals, in the mouth, or in the rectum. The draining lymph nodes become enlarged and hard, though painless. This resolves spontaneously over a few weeks.

Between three and six weeks after the chancre has appeared, the secondary stage begins, which can take consist of papules, macules, pustules, and even small fleshy lumps called condylomata lata. There may also be patches of ulceration on the mucous membranes (snail-track ulcers). The patient is still infectious. The lesions resolve within several weeks and the patient enters the latent phase, but at some time in the future will develop tertiary syphilis.

Tertiary syphilis presents with punched-out skin lesions called gumma (the most common form), damage to the central nervous system, or cardiovascular disease, which usually takes the form of aortic damage that can lead to aneurysms, valvular damage, or coronary artery damage, usually between 30 and 40 years after the primary lesion. The neurological manifestations are paralysis and a condition called tabes dorsalis (loss of parts of the spinal cord) in which balance is lost.

TREATMENT

Penicillin given intramuscularly for two weeks has revolutionized treatment. However, penicillin is of no value in tertiary syphilis.

PREVENTION

No vaccine is available. Safe sex, including using condoms, is the mainstay of prevention.

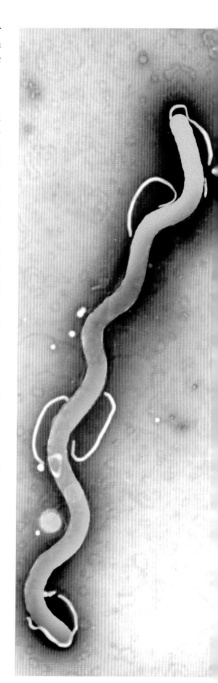

bacteroides fragilis/wound infections

■ **ORIGIN**
A bacterium normally residing in the human colon.

■ **FIRST IDENTIFIED**
In 1898.

■ **PORTAL OF ENTRY**
Part of the normal flora.

■ **INCUBATION PERIOD**
Some 2–3 weeks post colonic surgery.

Since it will not grow in, and is even destroyed by, even the smallest amount of oxygen, *Bacteroides fragilis* is termed a strict anaerobe. It is part of the normal flora of the gastrointestinal tract and produces a wide array of virulence mechanisms, for example, that evade the immune system and release enzymes that break down protein (proteases).

However, it does not produce disease unless it leaves its colonic home and gains access to normally sterile tissues. This occurs, for example, when there is a perforation of the intestine or following a burst appendix. It can also occur after colonic surgery. The colonic bacteria then leak into the peritoneum, grow, and remove oxygen, allowing the *B. fragilis* to produce a large abscess.

CLINICAL FEATURES
The abscesses form around the appendix or colon, or in the pelvic region, causing fever and abdominal pain. Postsurgery, they can even cause the sutures holding the surgical incision together to burst.

TREATMENT
B. fragilis is sensitive to metronidazole.

PREVENTION
No vaccine is available, nor likely. Metronidazole is given prior to operating to decrease the risk of infection following colonic surgery.

fusobacterium necrophorum/
necrobacillosis

Fusobacterium necrophorum is an anaerobic bacterium and is one of a number of other fusobacterium species that are part of the normal oral flora, the most numerous of which is *F. nucleatum,* which can cause periodontal disease.

    F. necrophorum is one of the few anaerobes that is a primary pathogen, that is, able to cause disease without prior tissue damage or immune suppression. It is a minority constituent of the normal oral flora that produces a number of virulence mechanisms including a hemolysin and the ability to survive inside macrophages.

CLINICAL FEATURES
Necrobacillosis begins with a sore throat, fever, and chills. The tonsils are covered with a membrane and the breath is foul-smelling (anaerobes produce volatile fecal-smelling gases). If the infection is not treated, the bacterium disseminates around the bloodstream. It produces abscesses where it lodges in, for example, the liver, brain, and spleen. It can also be fatal.

TREATMENT
Penicillin.

PREVENTION
No vaccine or other preventative is available. Early treatment can prevent fatality.

QUICK REFERENCE

■ **ORIGIN**
A bacterium present in a number of animal species, including humans.

■ **FIRST IDENTIFIED**
F. necrophorum was first isolated in 1884 from a case of calf "diphtheria."

■ **TRANSMISSION**
Unknown.

■ **PORTAL OF ENTRY**
Tonsils.

■ **INCUBATION PERIOD**
Unknown.

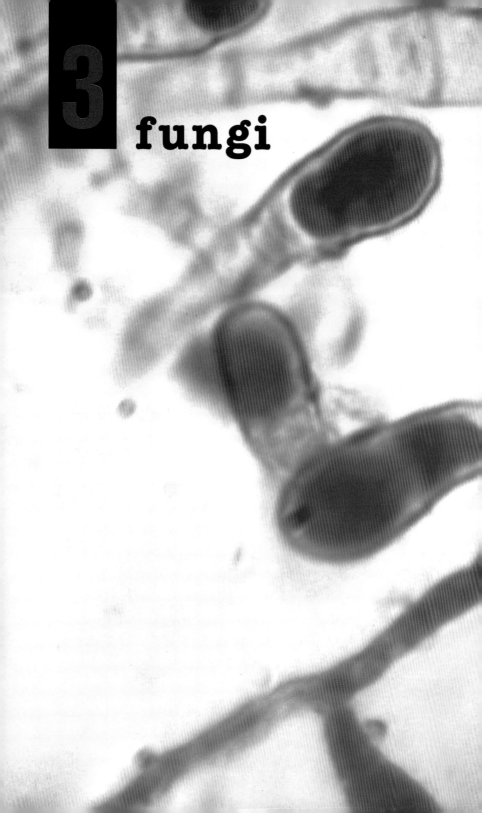

3 fungi

The fungi form a very large kingdom but only a small number are human pathogens. Fungi have chromosomes in a nucleus as well as internal organelles. They are usually larger than bacteria and reproduce both sexually and by binary fission. Some remain as single cells (yeast), whereas others form large branching networks. They can be subdivided into those that cause superficial infections or infections in the tissues beneath the skin, and those that disseminate around the body.

epidermophyton floccosum/
ringworm

This is one of a group of fungi that are adapted to gaining nutrients from keratinized tissue (skin, hair, and nails). They are collectively known as dermatophytes. *Epidermophyton floccosum* produces mycelia in skin. The mycelia consist of long fibrils burrowing just beneath the epidermis, which become segmented. They are quite hardy and shed with skin to infect the next patient. *E. floccosum* is anthropophilic, that is, human-specific.

CLINICAL FEATURES
The infections caused by dermatophytes are commonly called tinea. *E. floccosum* causes tinea cruris (ringworm of the groin or athlete's crutch), tinea pedis (athlete's foot), and tinea unguium (infection of the nails).

Tinea cruris begins with an itchy rash with a raised border extending from the groin down the upper thigh. It occurs particularly in the tropics and in soldiers or prisoners. The raised inflamed edge is the battleground between the advancing fungus and host defenses as it moves out radially from the origin of infection. Originally, it was thought that this raised, red-colored edge was the result of a long circular worm, hence the name ringworm.

Tinea unguium or onychomycosis is invasion of the nail plate, predominantly of the toes. It is estimated to affect between 2 percent and 4 percent of the population. The affected nails become thickened and opaque and eventually the nail can be lost.

TREATMENT
For tinea cruris and tinea pedis, topical application of antifungals such as clotrimazole or miconazole work well. The treatment of tinea unguim requires prolonged (initially three months) use of turbinafine or itraconazole.

PREVENTION
Infection can be very difficult to prevent but good personal hygiene is essential.

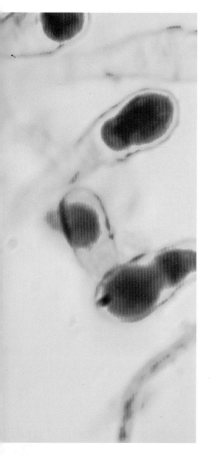

trichophyton spp/
ringworm

There is a large number of different *Tricho-phyton* species that can infect humans. They are usually subdivided into anthropophiles (transmitted from human to human), zoophiles (transmitted from animal to human), and geophiles (acquired from the environment).

Trichophyton can cause tinea capitis (scalp ringworm), tinea corporis (body ringworm), tinea barbae (ringworm of the beard area, or barber's itch), tinea cruris (athlete's crutch), tinea pedis (athlete's foot), and tinea unguium (nail ringworm).

In general, zoophilic infection produces a much more florid inflammation. Instead of appearing as a classical ringworm, it forms a large reddened lump sometimes called a kerion.

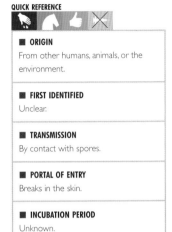

QUICK REFERENCE

■ **ORIGIN**
From other humans, animals, or the environment.

■ **FIRST IDENTIFIED**
Unclear.

■ **TRANSMISSION**
By contact with spores.

■ **PORTAL OF ENTRY**
Breaks in the skin.

■ **INCUBATION PERIOD**
Unknown.

CLINICAL FEATURES

Athlete's foot is the commonest form of fungus infection worldwide. It affects up to 65 percent of the population, and is more common in males and in hot, moist climates. The lesions usually begin in the webs between the toes. These are extremely itchy and large breaks in the skin occur.

Barber's itch occurs on the bearded area of the face and neck. It can present as one or two boggy lesions near the angle of the jaw or as a diffuse pustular folliculitis. The hairs are brittle and lusterless.

Body ringworm is a scaly, itchy rash affecting the trunk or proximal limbs.

TREATMENT

Terbinafine, itraconazole, or griseofulvin is given orally. Topical treatment with clotrima-zole or miconazole can be effective.

PREVENTION

No vaccine is available.

microsporum spp/ringworm

■ **ORIGIN**
Fungi can be transmitted from humans, animals, or the inanimate environment.

■ **FIRST IDENTIFIED**
Unclear.

■ **TRANSMISSION**
By direct or indirect contact, or airborne.

■ **PORTAL OF ENTRY**
Through the skin.

■ **INCUBATION PERIOD**
Unknown.

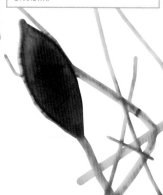

This third group of dermatophytic fungi include anthropophilic, zoophilic, and geophilic species (see page 165). They are a cause of tinta capitis (scalp ringworm), tinea corporis (body ringworm), and tinea barbae (barber's itch).

The different *Microsporum* species are the most frequent causes of tinea capitis, which mainly affects boys of school age. Tinea capitis is highly contagious and occurs in epidemics. The vehicles for transmission appear to be barber's clippers and cinema seats, although sharing combs can also spread infection, which begins as a small, itchy, scaly papule. This spreads radially and hairs in the advancing circular area become brittle and break off a few millimeters above the scalp, leaving irregular, circular grayish areas with hair stumps.

CLINICAL FEATURES
As for *E. floccosum* and *Trichophyton spp* (see pages 164–165).

TREATMENT
Orally, with terbinafine, itraconazole, or griseofulvin. Topically, with clotrimazole or miconazole.

PREVENTION
No vaccine is available.

malessezia spp/pityriasis versicolor, seborrheic dermatitis

Malessezia form a group of yeasts that grow in the skin's relatively fatty (lipophilic) sebaceous secretions. They are unicellular fungi and can be found as part of the normal skin flora.

The most important species are *M. furfur*, *M. dermatis*, *M. globus,* and *M. sympodialis*, which have long been associated with pityriasis versicolor, though recent work has demonstrated that they can also cause seborrheic dermatitis (dandruff) in susceptible individuals. It is our immune response to the fungus that produces seborrheic dermatitis.

CLINICAL FEATURES

Pityriasis versicolor is a chronic, relatively benign skin condition characterized by scaly hypo- or hyperpigmented flat lesions (macules) that mainly affect the parts of the body most richly endowed with sebaceous glands (the neck, upper trunk, or upper arms). The lesions may coalesce to produce large areas of disease.

Seborrheic dermatitis occurs on the face, chest, and scalp. Dandruff, which affects between 5 percent and 10 percent of the population, is the mildest form of seborrheic dermatitis. Disease occurs more frequently in the winter months.

TREATMENT

For pityriasis versicolor, options are selenium (found in shampoos) and oral antifungals (ketoconazole or itraconazole), used for between one and two weeks. For seborrheic dermatitis, which is a chronic condition, long-term (four weeks) treatment with oral antifungals is required. Topical treatment with hydroxypyridones (for example, ciclopirox) in shampoo is usually effective.

PREVENTION

Itraconazole given once a month for six months can prevent recurrence of pityriasis versicolor. There is no proven prevention for seborrheic dermatitis.

QUICK REFERENCE

■ **ORIGIN**
A normal fungal commensal.

■ **FIRST IDENTIFIED**
Link between *Malessezia* and seborrheic dermatitis realized only in the last five years.

■ **TRANSMISSION**
Unknown.

■ **PORTAL OF ENTRY**
The skin.

■ **INCUBATION PERIOD**
Not applicable.

sporothrix schenckii/
sporotrichosis

QUICK REFERENCE

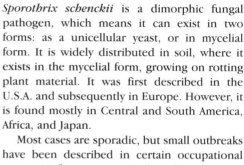

■ **ORIGIN**
A soil saprophyte that accidentally infects humans.

■ **FIRST IDENTIFIED**
In Brazil in 1908.

■ **TRANSMISSION**
From soil.

■ **PORTAL OF ENTRY**
Through the skin.

■ **INCUBATION PERIOD**
1–10 weeks.

Sporothrix schenckii is a dimorphic fungal pathogen, which means it can exist in two forms: as a unicellular yeast, or in mycelial form. It is widely distributed in soil, where it exists in the mycelial form, growing on rotting plant material. It was first described in the U.S.A. and subsequently in Europe. However, it is found mostly in Central and South America, Africa, and Japan.

Most cases are sporadic, but small outbreaks have been described in certain occupational groups. Infection occurs when fungal spores are introduced into the skin by a penetrating wound (for example, from thorns or splinters). Once inside the body, the fungus reverts to the yeast-like form.

CLINICAL FEATURES
Between about 1 and 10 weeks after the spores are introduced into the skin, a reddish-purple necrotic lesion appears, around which satellite lesions may develop. The fungus then spreads along the draining lymphatic vessels leaving trails of secondary lesions. Spontaneous healing can occur but this is rare. Without treatment, the lesions persist for years, progressing slowly and leaving scarred and disfigured skin.

TREATMENT
A saturated potassium iodide solution given orally for between two and four months is a very effective though unpleasant therapy. Antifungals such as itraconazole are also useful.

PREVENTION
Avoiding contact with *Sporothrix schenckii* is the only way to prevent sporotrichosis.

phialophora verrucosa/
chromoblastomycosis

Phialophora verrucosa is one of a group of pigmented fungi that cause chromoblastomycosis (*chromos* is color in Greek). The disease occurs most commonly in tropical regions but there have also been incidences in North America, Europe, the U.K., and South Africa.

Ph. verrucosa is a saprophyte living on decaying vegetable matter in soil. Humans most often become infected when walking barefoot on soil in which the fungus is growing. It gets into the body via any puncture wounds on the feet. It cannot be spread person-to-person. *Ph. verrucosa* is the commonest cause of chromoblastomycosis in the U.S.A., whereas worldwide it is most often caused by *Fonsecaea pedresoi*.

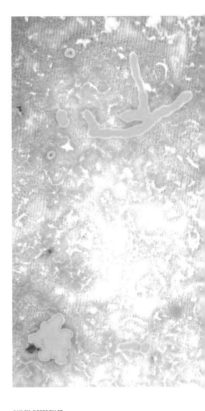

CLINICAL FEATURES
Disease begins with a small pink nodule, or group of nodules, in the wound area. The nodules coalesce and become warty and pigmented. They can break down and may have an offensive smell. They persist and gradually extend, sometimes causing severe limb deformity. After many years this can lead to skin cancer (squamous carcinoma).

TREATMENT
Antifungals used most often are turbinafine, itraconazole, or flucytosine (usually in combination), heat treatment to the area (the fungus is not well adapted to temperatures over 99°F/37°C), or cryotherapy.

Surgical removal is used only if the lesion is small and in the early stages, as removing large, older lesions can lead to regrowth in the scar.

PREVENTION
Preventing skin wounds coming into contact with infected soil.

QUICK REFERENCE

■ **ORIGIN**
A saprophyte living in soil.

■ **FIRST IDENTIFIED**
In the 1950s.

■ **TRANSMISSION**
Via penetrating wounds contaminated by soil containing *Ph. verrucosa*.

■ **PORTAL OF ENTRY**
Through the skin.

■ **INCUBATION PERIOD**
Weeks to months.

rhinosporidium seeberi/
rhinosporidiosis

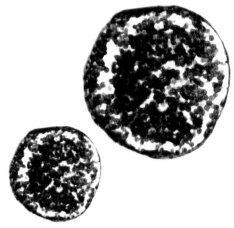

■ **ORIGIN**
Unknown.

■ **FIRST IDENTIFIED**
In 1950. Assigned to the protista in 1999.

■ **TRANSMISSION**
Unknown.

■ **PORTAL OF ENTRY**
Mucous membranes of eyes, nose, and larynx.

■ **INCUBATION PERIOD**
Unknown.

Rhinosporidium seeberi causes the rare disease rhinosporidiosis, which is most common in India and Sri Lanka. Although it has been classified as a fungus for decades, recent work suggests it is not a fungus but a member of a newly described kingdom of microorganisms called the protista that have so far been found only in fish.

CLINICAL FEATURES
R. seeberi causes a localized infection of the mucous membranes, usually of the nose or conjunctivae. It forms a friable (disintegrates easily) polyp, which can be large and grape-like and can persist for long periods.

TREATMENT
Antimicrobial therapy is ineffective. The only effective treatment is surgical removal though this does not prevent recurrence.

PREVENTION
No vaccine is available.

basidiobolus ranarum/
subcutaneous zygomycosis

Basidiobolus ranarum is a mold-like fungus found in the environment. It does not grow well at temperatures below 60°F (15°C) and so is rarely a cause of disease in temperate countries.

In tropical and subtropical climates, in particular east and west Africa, Brazil, India, and Indonesia, it causes two infections. The more common of the two (although still rare) is subcutaneous zygomycosis. This tends to occur in adolescent or young adult males and is transmitted through skin wounds from soil contaminated with the fungus.

The second infection, gastrointestinal basidobolomycosis, is very rare and there have been only six cases worldwide—in Nigeria, Brazil, Kuwait, and the U.S.A. (predominantly in Arizona). It is probable that infection is acquired by ingestion of soil, animal feces, or food contaminated by either of these.

CLINICAL FEATURES
Subcutaneous zygomycosis usually occurs at the site of infection weeks to months after the skin wound. It presents as a deforming swelling with a woody consistency. Gastrointestinal basidobolomycosis presents with abdominal pain that has been mistaken for carcinoma, Crohn's disease, and diverticular disease. Infection most often occurs in the colon.

TREATMENT
Itraconazole and terbinafine are used but surgical removal of the subcutaneous or gastrointestinal lesions is necessary.

PREVENTION
No information is currently available.

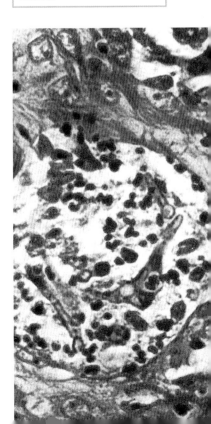

streptomyces somaliensis/
mycetoma, madura foot

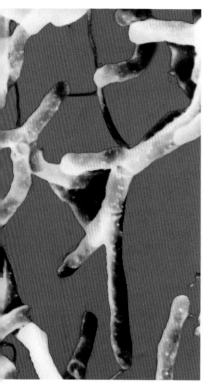

Mycetoma is a localized chronic infection that affects the skin, subcutaneous tissues, and underlying bone. It is caused by a variety of bacteria (*Actinomyces, Streptomyces,* and *Nocardia*) and fungi (*Madurella mycetomatis, Fusarium, Aspergillus nidulans,* or *Scedosporium*). All are saprophytes living in the soil. Infection is acquired, via skin wounds, from soil or other material contaminated by the bacteria or fungi.

Mycetoma is generally a disease found in the tropics or subtropics, particularly in areas of low rainfall, for example in parts of Africa (Sudan, Somalia), Mexico, the central and northern parts of South America, the Middle East, and India. *Nocardia* bacteria are usually responsible for cases of mycetoma in Central America, and *Madurella mycetomatis* in Africa and India.

The disease is more common in males than females, in adults, and in agricultural workers. After penetrating the skin, the bacteria or fungi grow, extending deeper into the bone and bursting to produce sinuses that drain the purulent discharge.

CLINICAL FEATURES
The incubation period is unknown as patients usually have no recollection of the injury that caused the infection. Mycetoma begins as a small, painless swelling in the skin or subcutaneous tissue. This slowly enlarges and sinuses form, draining the pus. The swelling is painful and woody at this stage. There is further deformity as the lesion progresses locally.

TREATMENT
Actinomyces or *Streptomyces* bacterial infections respond to cotrimoxazole. Treatment of the fungi is much more difficult. Ketoconazole, itraconazole, or griseofulvin are possibilities.

PREVENTION
Avoiding skin wounds.

QUICK REFERENCE

■ **ORIGIN**
Soil microorganisms.

■ **FIRST IDENTIFIED**
Clinical descriptions are from the early 1900s.

■ **TRANSMISSION**
From soil via penetrating wounds such as splinters.

■ **PORTAL OF ENTRY**
Through the skin.

■ **INCUBATION PERIOD**
Unknown.

histoplasma capsulatum/
histoplasmosis

Histoplasma capsulatum is a dimorphic fungus. It grows as a mold in soil, but as a yeast in human tissues. It is found in particularly high concentrations in soil beneath bird or bat roosts, and it is therefore thought that bird and bat feces provide the nutrients for the fungus to grow.

Histoplasmosis is endemic in parts of the U.S.A., in the West Indies, Central and South America, Africa, India, and the Far East. Human cases are usually sporadic, and infection is acquired by inhalation. At particular risk are those who are in contact with bats or birds in confined spaces, for example, chicken farmers.

CLINICAL FEATURES
Most cases are asymptomatic, as only one in a hundred of those exposed actually develops disease. The acute febrile illness, with a cough and joint and chest pains, does not usually require treatment. However, in some patients there is chronic progressive pulmonary disease. In a minority of cases, histoplasmosis disseminates around the body and can prove fatal.

TREATMENT
Treatment is not usually necessary.

PREVENTION
Avoiding exposure to bird and bat feces.

QUICK REFERENCE

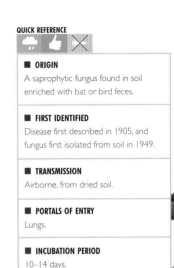

- **ORIGIN**
  A saprophytic fungus found in soil enriched with bat or bird feces.

- **FIRST IDENTIFIED**
  Disease first described in 1905, and fungus first isolated from soil in 1949.

- **TRANSMISSION**
  Airborne, from dried soil.

- **PORTALS OF ENTRY**
  Lungs.

- **INCUBATION PERIOD**
  10–14 days.

coccidioides immitis/
coccidioidomycosis

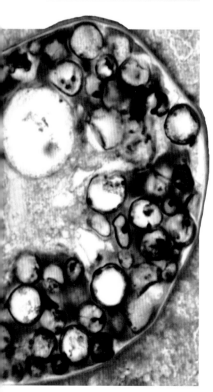

Coccidioides immitis is a dimorphic fungus. In its saprophytic form it exists in soil as a branching mycelium. It exists in tissues as a spherule (a small sphere), inside which are large numbers of endospores that are released when the spherule ruptures.

The fungus is transmitted to humans by inhalation of fragments of the mycelium (called arthrospores). Those who are likely to become infected are archeologists, fieldworkers, and anyone working where topsoil is dry then disturbed so as to allow the fungus to become airborne. So in endemic regions, outbreaks of coccidioidomycosis often follow earthquakes.

Coccidioidomycosis is endemic in southwest U.S.A. (California, Arizona), northern Mexico, Guatemala, Honduras, Venezuela, Colombia, Paraguay, Bolivia, and Argentina, where each year between 3 percent and 5 percent of the population become infected. Epidemics can occur when dust storms carry arthrospores to more populated areas.

CLINICAL FEATURES
Only 40 percent of those infected develop disease, most of whom present with an acute flu-like illness that resolves by itself. Around 5 percent will be left with cavities or shadows in their lungs. In a minority of cases, the fungus spreads around the body via the bloodstream, may lodge in the skin and produce large warty lesions or subcutaneous abscesses. It can also produce chronic meningitis, osteomyelitis (infection of the bone), and peritonitis.

TREATMENT
Most infections do not require treatment. Severe disseminated infection is treated with intravenous amphotericin B.

PREVENTION
No vaccine is available. Avoidance of the arthrospores or activities likely to generate airborne spores are the only prevention.

cryptococcus neoformans/
cryptococcosis

Cryptococcus neoformans is another dimorphic fungus existing in mycelial form in the environment and as a yeast when it infects humans. There are two varieties of *C. neoformans*: var. *neoformans* and var. *gattii*. The mycelial form of *C. neoformans* var. *gattii* is found in eucalyptus trees, and *C. neoformans* var. *neoformans* has been found in the droppings of pigeons and other birds, and a variety of fruits.

It has been known to be a human pathogen since 1894. However, it has only been since the emergence of the AIDS pandemic that large numbers of infections have occurred. It is transmitted to humans (for example, pigeon fanciers) from environmental sources via inhalation, though usually only those whose immune system is compromised by AIDS, or immunosuppressive cytotoxic drugs, develop disease.

CLINICAL FEATURES
Although *C. neoformans* var. *gattii* can cause lung infection in otherwise healthy individuals, it is rarely reported. *C. neoformans* var. *neoformans* causes pneumonia, meningitis, or both, in immunocompromised patients.

TREATMENT
Intravenous amphotericin B and flucytosine are the standard treatments. The disease is curative in up to three-quarters of non-AIDS patients and thus antiretroviral drugs are the best therapy.

PREVENTION
Suppressive therapy with fluconazole after an episode of cryptococcal infection in AIDS patients is successful until the onset of resistance.

QUICK REFERENCE

■ **ORIGIN**
An environmental fungus.

■ **FIRST IDENTIFIED**
In 1894.

■ **TRANSMISSION**
From the environment by inhalation. No person-to-person spread.

■ **PORTAL OF ENTRY**
Inhalation into the lungs.

■ **INCUBATION PERIOD**
Not clear.

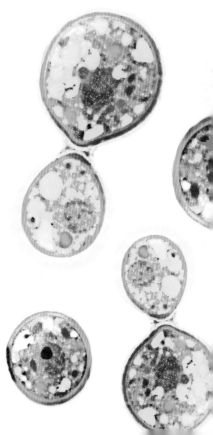

candida spp/candidosis

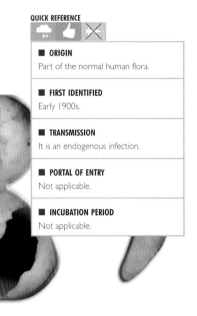

■ **ORIGIN**
Part of the normal human flora.

■ **FIRST IDENTIFIED**
Early 1900s.

■ **TRANSMISSION**
It is an endogenous infection.

■ **PORTAL OF ENTRY**
Not applicable.

■ **INCUBATION PERIOD**
Not applicable.

Of the different *Candida* species, *C. albicans* is the most common human pathogen. It is a yeast that does not form mycelia, and is widely distributed in soils, food, water, fruit-juice drinks, and plants. It is part of our normal flora and infection is therefore most often acquired from within.

The infection is usually confined to the skin and mucous membranes. However, in premature babies, AIDS patients, or those being treated with cytotoxic or immunosuppressive drugs, for example, infection can spread throughout the body. Administration of broad-spectrum antibiotics increases the chances of disease.

CLINICAL FEATURES

Infection is of the nails, skin around the nails, or areas of skin that touch each other and remain moist, such as the groin or beneath the breasts, where it causes intense inflammation. It may cause demarcated white patches in the mouth (oral thrush) or vagina (vaginal thrush).

In immunocompromised patients, infection can spread from the mouth to produce enormous white plaques covering the oesophagus or bronchi. It can disseminate in the blood to the meninges, eyes, or heart, or ascend the urethra to cause urinary tract infection.

TREATMENT

Mucocutaneous infection usually responds to topical therapy with clotrimazole. Occasionally, it is necessary to give oral terbinafine, fluconazole, or itraconazole. For disseminated infection, amphotericin B is given intravenously.

PREVENTION

Long-term fluconazole treatment is given to AIDS patients but this can lead to replacement of *C. albicans* with other, resistant, *Candida* species.

pneumocystis jirovecii/pneumocystosis

Until the early 1990s, *Pneumocystis jirovecii* was thought to be a protozoan, based on its morphology and response to antiprotozoal drugs. Using molecular biological tools, it is now clear that it is really a fungus. Its name is in honor of the doctor who described large epidemics of disease seen in malnourished children in Eastern Europe after World War II.

It is found worldwide and it seems that many of us become infected in childhood but do not develop disease. Disease occurs only if the immune system is unable to cope with infection, for example due to malnutrition, leukemia, or AIDS. It is thought that in some cases, infection is due to *P. jirovecii* acquired in early life that has persisted and reemerges when immunity is damaged.

Airborne transmission means the disease can be spread person-to-person. The fungus enters the alveoli of the lungs where it grows and initiates the disease process.

CLINICAL FEATURES
Most often, *P. jirovecii* causes lung infection or pneumonitis. It presents with rapid breathing, difficulty breathing, fever, and cough. This worsens gradually and, if untreated, will eventually lead to death.

TREATMENT
High-dose cotrimoxazole. Other drugs include pentamidine or dapsone. Antifungal drugs have no effect.

PREVENTION
No vaccine is available. Once an AIDS patient has had one episode of *P. jirovecii* pneumonia they are at much greater risk of a second, so prophylaxis with cotrimoxazole or pentamidine is given. However, antiretroviral drugs have had a great effect, such that *P. jirovecii* pneumonia is now rare.

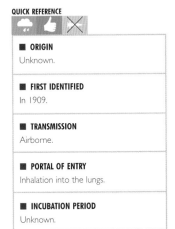

QUICK REFERENCE

■ **ORIGIN**
Unknown.

■ **FIRST IDENTIFIED**
In 1909.

■ **TRANSMISSION**
Airborne.

■ **PORTAL OF ENTRY**
Inhalation into the lungs.

■ **INCUBATION PERIOD**
Unknown.

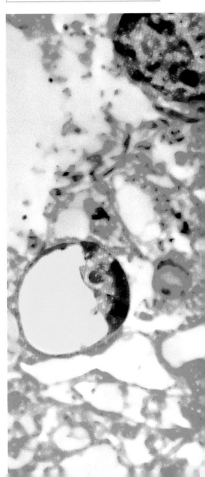

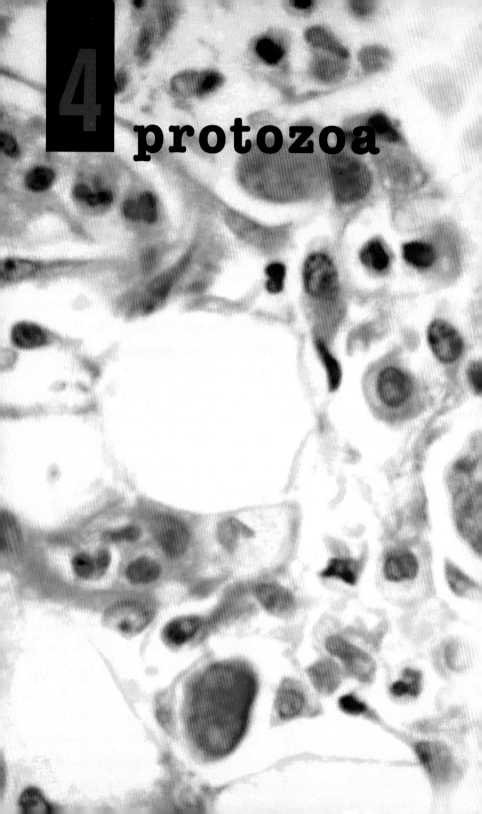

4 protozoa

The protozoa form a large kingdom of eukaryotes consisting of single-celled organisms. They have nuclei and intracellular organelles but do not have a cell wall. Some reproduce by binary fission and others by sexual reproduction. They are quite fragile, and those that are shed into the external environment usually produce hardy thick-walled cysts. Only a minority of the protozoa are human pathogens.

cryptosporidium species/
cryptosporidiosis

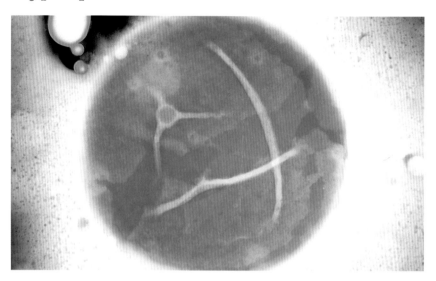

There are several different *Cryptosporidium* species that infect humans. *C. hominis* is solely a human pathogen, whereas *C. parvum* (cattle, sheep, and horses), *C. meleagridis* (fowl), *C. muris* (mice), *C. canis* (dogs), *C. felis* (cats), and *C. wrairii* (guinea pigs) are acquired as a zoonosis but can also be transmitted from person-to-person.

C. muris was the first to be detected, in the gastric glands of mice, in 1907. However, it was not until 1976 that the first human infections were described. This is not to say that humans had not been infected previously, but that the protozoan had merely been overlooked. With the AIDS epidemic and development of simple diagnostic tools, its importance has now been realized.

TRANSMISSION
There are two major stages of cryptosporidium transmission. Oocysts (thick-walled eggs) that contain four sporozoites (living organisms) are excreted in feces in large numbers. The oocysts are relatively small but very tough. For example, they are not destroyed by chlorinating

water or by other disinfectants. Because of their small size, the oocysts can also get through the filtration systems that process our drinking water. The largest reported water-borne epidemic of cryptosporidiosis occurred in Milwaukee, U.S.A., when 403,000 people were infected. Person-to-person and food-borne outbreaks also occur.

Once ingested, the oocyst opens to release the sporozoites which eventually develop into oocysts that are excreted in feces to infect the next host.

CLINICAL FEATURES

Cryptosporidiosis is a particularly common cause of diarrheal disease in children and adults. The diarrhea persists for an average of 10 to 14 days but can last for months. The stool is watery green and very foul-smelling.

In infants, it can cause stunted growth if it persists for too long. In individuals whose immune systems are compromised, it produces a prolonged, life-threatening diarrhea.

TREATMENT

The only drug shown to have effect is nitazoxanide, which is not licensed in many countries.

PREVENTION

No vaccine is available. Boiling or freezing drinking water will kill the oocysts.

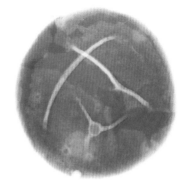
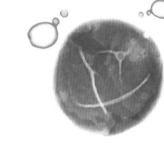

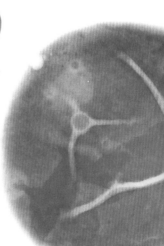
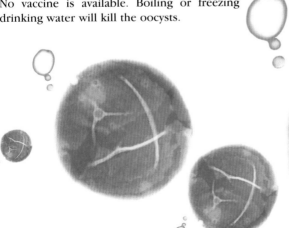

entamoeba histolytica/
amebic dysentery

QUICK REFERENCE

■ **ORIGIN**
A pathogenic protozoan of humans.

■ **FIRST IDENTIFIED**
In 1875.

■ **TRANSMISSION**
Feco-orally directly, or in food and water.

■ **PORTAL OF ENTRY**
Mouth to the intestine.

■ **INCUBATION PERIOD**
5–10 days.

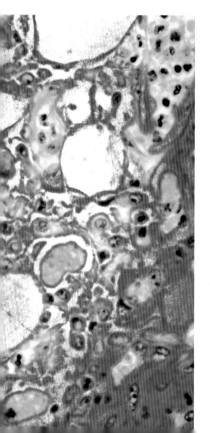

Entamoeba histolytica is a protozoan parasite excreted in the feces of infected persons as cysts. It was first described by Lösch in 1875 who observed the trophozoites (living amebae) that had ingested human red cells.

Ent. histolytica causes the death of the cells lining the colon, which elicits an inflammatory response. This produces a blood diarrhea (dysentery). It is estimated to cause some 100,000 deaths each year worldwide, the majority of which occur in the developing world, in particular tropical and subtropical regions. Cases in developed countries are usually as a result of travelers returning from endemic regions.

Infected patients excrete cysts. On hands, these are destroyed by drying within 10 minutes but survive for 45 minutes or longer in fecal material lodged under fingernails. In water they survive for up to 120 hours at 77°F (25°C), for months at 34–41°F (1–5°C), but are killed by boiling or pasteurization. They are relatively resistant to chlorination. The cysts are transmitted person-to-person feco-orally directly, or indirectly in food and water.

CLINICAL FEATURES
Amebic dysentery, in contrast to *Shigella* dysentery, is usually a "walking dysentery," that is, patients are not acutely ill. It produces chronic bloody diarrhea. However, some patients develop fulminating colitis, amebic appendicitis, or even hepatic abscesses (the contents of which are described as resembling anchovy paste).

TREATMENT
Metronidazole.

PREVENTION
Good personal hygiene and safe (not contaminated by human feces) food and water.

giardia intestinalis/
giardiasis

Giardiasis is the most common human infection caused by protozoa worldwide. It is found in all regions of the world—temperate, tropical, and subtropical. *Giardia intestinalis* (also known as *G. lamblia*) is a free-living intestinal parasite that causes diarrhea. It is excreted in feces as a cyst that can survive for months in cool, moist conditions and is relatively resistant to chlorination so, like *Cryptosporidium* species, it can produce waterborne epidemics. However, this is less likely in developed countries as the cysts are large and do not get through the potable-water filtration systems.

Infection follows ingestion of as few as 10 to 100 cysts. During infection, anything up to 20,000 cysts are excreted per gram of feces. Infection is acquired by ingestion of food or water contaminated by cysts, or even directly, for example in crèches where up to half of all children can be infected at any one time. Trophozoites emerge from the cysts and attach to the small intestinal mucosa. The cause of the diarrhea is unclear.

QUICK REFERENCE

■ **ORIGIN**
A pathogen of humans. Other animals have their own *Giardia* bacteria but these do not infect humans.

■ **FIRST IDENTIFIED**
The inventor of the microscope, Antonie von Leeuwenhoek, was the first to observe *G. intestinalis* in his own diarrheal stools in 1675. It was rediscovered by William Lamble in 1859, and first cultured in 1970.

■ **TRANSMISSION**
Feco-orally directly or in food and water.

■ **PORTAL OF ENTRY**
Mouth to the upper intestine.

■ **INCUBATION PERIOD**
3–20 days.

CLINICAL FEATURES
The most common form of giardiasis is asymptomatic infection, but it can also produce acute or chronic diarrheal disease. In most patients, this is a self-limiting illness with between 60 percent and 70 percent of individuals experiencing weight loss. It resolves within two to four weeks. However, between 30 percent and 50 percent of patients develop chronic diarrhea, losing up to a fifth of their body weight.

TREATMENT
Metronidazole.

PREVENTION
Prevention measures include good personal hygiene and safe drinking water and food.

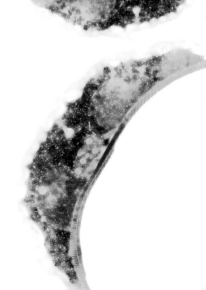

trypanosoma spp/
trypanosomiasis

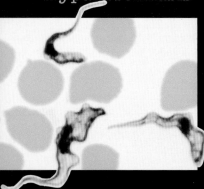

Trypanosoma are protozoa with a two-stage life cycle. The trypanosomes that infect humans cause African trypanosomiasis or sleeping sickness (*T. brucei gambiense* and *T. brucei rhodesiense*) and American trypanosomiasis or Chagas' disease (*T. cruzi*).

Sleeping sickness is predominantly a disease of game and cattle that is spread to humans by the blood-sucking tsetse fly. Chagas' disease is spread to humans by kissing-bugs, from its natural reservoir, the armadillo, or from multiplier hosts such as cats and dogs.

QUICK REFERENCE

■ **ORIGIN**
Pathogens of animals accidentally transmitted to humans.

■ **FIRST IDENTIFIED**
T. brucei rhodesiense in 1896, T. brucei gambiense in 1902, and T. cruzi in 1908.

■ **TRANSMISSION**
African trypanosomiasis by the tsetse fly, and American trypanosomiasis by kissing-bugs.

■ **PORTAL OF ENTRY**
Through the skin from insect bites.

■ **INCUBATION PERIOD**
African trypanosomiasis: 2–4 days.
American trypanosomiasis: 4–12 days.

CLINICAL FEATURES

Within a few days of the bite of an infected tsetse fly, the "trypanosomal chancre" develops, a lump that enlarges up to one inch (2.5 cm), then subsides after two or three weeks, with scarring. However, the protozoa will already have disseminated to lymph nodes and more importantly, to the brain. Meningoencephalitis occurs in all, and patients present with severe headache, sleep disorders, apathy, irritability, then paranoia, delusion, and mania, and eventually coma and death.

Initially, Chagas' disease is often inapparent. Three to 12 days after the insect bite there can be a painless red or violet nodule and the draining lymph nodes enlarge. It usually resolves within a couple of months. However, *T. cruzi* will have spread around the body. The chronic stage begins two to four months later. It affects the heart, which enlarges and malfunctions, leading to death in 5 percent of patients.

TREATMENT

For sleeping sickness, treatment involves suramin or pentamidine. Chagas' disease is treated with nifurtimox or benznidazole.

PREVENTION

No vaccines are available. Prevention is by controlling the disease in its reservoir hosts and the use of insecticides.

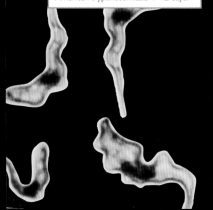

leishmania spp/kala azar, cutaneous leishmaniasis, mucocutaneous leishmaniasis

Leishmania are protozoan parasites transmitted by sandflies. Visceral leishmaniasis (Kala Azar) is due to *L. infantum* or *L. donovani* (depending on the country). The main reservoir host for *L. infantum* is the dog. Sandflies become infected by feeding on infected dogs.

Old World cutaneous leishmaniasis (Oriental sore) is due to *L. major, L. tropica,* and *L. aethiopica.* The variety of reservoir hosts includes the great gerbil, fat sand rat, Nile rats, and humans. The disease can cause fatal epidemics.

New World leishmaniases are mucocutaneous lesions due to *L. mexicana* (chicolero ulcer), *L. braziliensis* (espundia), *L. panamensis,* and *L. guyanensis,* and have a variety of rodent hosts.

CLINICAL FEATURES

L. donovanis and *L. infantum* cause a small ulcerated lump or leishmanioma that may heal or progress to Kala Azar, which is characterized by fever, enlarged spleen and liver, or anemia. Secondary infections with bacteria and hemorrhages are the main complications, which occasionally lead to death if untreated.

Cutaneous leishmaniasis presents as a small, red, inflamed lump at the site of the insect bite (usually on the face). This enlarges, forming a painless nodule with a surface crust. Eventually it heals by scarring, but in a minority of cases progresses to chronic cutaneous leishmaniasis in which the ulcers can be very disfiguring.

Chicolero and espundia present as a painless, itchy papule that enlarges and ulcerates. In espundia, the primary cutaneous lesion extends to the mucosa, and leads to disintegration of the nasal cartilage.

TREATMENT

Pentavalent antimony components (meglumine antimoniate), pentamidine, ketoconazole, and amphotericin B. Surgery may be required.

PREVENTION

No vaccine is available.

QUICK REFERENCE

■ **ORIGIN**
Protozoa infecting a number of different reservoir hosts.

■ **FIRST IDENTIFIED**
The first description of Kala Azar was by Twining in 1832. Leishman described the pathogen in 1903 in a patient with Kala Azar. Cutaneous leishmaniasis is described in Assyrian texts from 650 B.C., and a water jug from the Peruvian Nazca period (200 B.C.–600 A.D.) appears to depict mucocutaneous leishmaniasis.

■ **TRANSMISSION**
From reservoir hosts (rodents, dogs, or humans) by sandflies.

■ **PORTAL OF ENTRY**
Through the skin via sandfly bites.

■ **INCUBATION PERIOD**
Kala Azar: 2–6 months.
Oriental sore: 2–3 weeks.
Mucocutaneous leishmaniasis: 1–2 months.

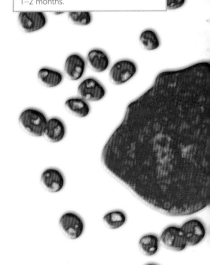

plasmodium falciparum/malaria

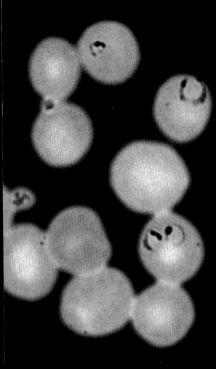

Plasmodium falciparum causes the most severe form of malaria. The disease has been written about since 490 B.C. when it was recognized that stagnant water resulted in epidemics of fever. However, long before this, native Peruvian tribes were aware that the bark of the Cinchona tree (which contains quinine) could be used to cure or prevent malaria. Transmission of the malaria parasite by anopheline mosquito bite was recognized in 1897.

It is estimated that, worldwide, there are up to 500 million cases of malaria each year, causing around 2.7 million deaths. Around a million of these are in children under five. Cases of malaria in Europe and North America are usually imported. In many African countries, cases of malaria peak in the rainy season because mosquitoes need stagnant water for the larval phase of their life cycle.

CLINICAL FEATURES
Patients develop a fever which may be continuous or remittent and accompanied by chills. Progression to severe disease can be rapid and is most common in children or adult travelers.

Severe disease includes cerebral malaria (fits and altered consciousness), acute kidney failure, or shock. Blackwater fever results from massive, rapid disruption of red cells, releasing hemoglobin which passes through the kidneys turning urine black or a dark reddish-brown.

TREATMENT
Quinine or artemether.

PREVENTION
Although trials of malaria vaccines are in progress, none are completed. Prevention is by insecticides, use of bed nets impregnated with insecticides, or giving prophylactic antimalarials such as chloroquine, proguanil, mefloquine (Larium), doxycycline, or malarone. Chemoprophylaxis is essential for Europeans or North Americans if traveling to malarious areas.

QUICK REFERENCE

■ **ORIGIN**
A human pathogen that has probably evolved with humans.

■ **FIRST IDENTIFIED**
In ancient literature. The protozoan was first described in 1886 and transmission by mosquitoes in 1897.

■ **TRANSMISSION**
By the bite of female anopheline mosquitoes, which are biological vectors.

■ **PORTAL OF ENTRY**
Through the skin, via the mosquito's proboscis.

■ **INCUBATION PERIOD**
9–14 days for falciparum malaria.

toxoplasma gondii/toxoplasmosis

Toxoplasma gondii is a protozoan parasite naturally maintained in a cat–rodent, predator–prey cycle, though sheep, pigs, and humans can become infected accidentally. Cats become infected by eating infected rodents. They are persistently infected and excrete oocysts (eggs containing living sporozoites) in their feces, which can persist for weeks or months in moist soil.

Pigs or sheep become infected by eating *T. gondii* oocysts in soil or bedding. Humans can become infected by accidentally eating oocysts in cat feces. For example, cleaning cat litter trays generates airborne oocysts, and cats often use children's sandpits as litter trays. Infection can also result from the ingestion of undercooked pork or lamb. Freezing meat destroys the bacteria.

CLINICAL FEATURES

In between 60 percent and 70 percent of cases, humans become infected asymptomatically. However, asymptomatic patients can become symptomatically infected if their immune systems then become suppressed. A glandular-fever-like illness with fever, a rash, and enlargement of the lymph nodes is the most common presentation.

Other forms of toxoplasmosis are congenital and recrudescence of persistent infection after immune suppression such as AIDS. In congenital toxoplasmosis, if a mother is infected for the first time with *T. gondii* during pregnancy, it can cross the placenta to infect the fetus. In many cases, the newborn is normal at birth but damage to the retina occurs later and can lead to blindness. In the most severe cases, there is microcephaly (a small head due to an underdeveloped brain) and mental retardation. In recrudescent toxoplasmosis, the most frequent manifestations are acute or chronic encephalitis, myocarditis, and myositis.

TREATMENT

A combination of two drugs: pyrimethamine and sulfadiazine.

PREVENTION

No vaccine is available for humans.

QUICK REFERENCE

■ **ORIGIN**
A parasite of cats and rodents accidentally transmitted to humans.

■ **FIRST IDENTIFIED**
T. gondii was first described in 1908 and human infections in 1937–39.

■ **TRANSMISSION**
By ingestion of oocysts from cat feces or bradyzoites in improperly cooked pork, lamb, or mutton. From mother to fetus across the placenta.

■ **PORTAL OF ENTRY**
Intestine.

■ **INCUBATION PERIOD**
5–18 days.

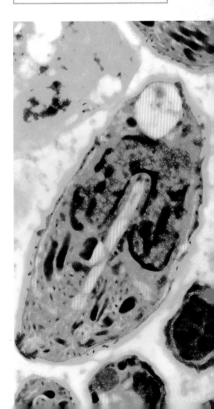

glossary

AEROSOL Suspension of particles in the air generated by coughing.

ANAEROBE An organism that can live in the absence of oxygen.

ANOREXIA Aversion to food.

ANTIBIOTIC An antibacterial drug derived from another microbe.

ANTIBODY Defense protein that recognizes invading microbes and helps to kill them.

ANTIGEN Complex foreign molecules on microbes recognized by antibodies.

ANTIGENIC DRIFT Gradual change in the antigens of influenza virus by mutation.

ANTIGENIC SHIFT Major change in the antigens of influenza virus so no one has a defense against it. Leads to pandemics.

ANTIMICROBIAL An antibacterial drug either chemically synthesized or antibiotic.

ANTIVIRAL A drug designed to combat viral infection.

ARTHRALGIA Pain in the joints.

ARTHRITIS Inflammation of the joints.

BACILLI Rod-shaped bacteria.

BACTEREMIA Literally, bacteria in the bloodstream.

BIOLOGICAL TRANSMISSION Insect transmission in which the microbe multiplies in the insect.

BRADYZOITE A slowly dividing form of the protozoan *Toxoplasma gondii.*

BRONCHIOLITIS An infection of infants with inflammation and blockage of the bronchioles in the lung.

BUBO A pus-filled swelling, often in the groin.

CARCINOMA A malignant tumor of skin or mucous membranes.

CARDIO- Relating to the heart.

CIRRHOSIS Chronic inflammation of the liver which leads to its destruction.

CLOTTING FACTOR A cascade of blood proteins that cause blood to clot.

COCCI Spherical-shaped bacteria.

COMMENSAL An organism in a symbiotic relationship that derives benefit while the other is unaffected.

CONFLUENT Refers to a rash where the spots are so numerous that they run together.

CONJUNCTIVITIS Inflammation of the conjunctivae, the membrane that covers the eye and inner surface of the eyelids.

CORYZA Runny nose.

CUTANEOUS Relating to the skin.

CYTOKINES Small molecules that mediate inflammatory responses by cells.

CYTOTOXIC DRUGS Anticancer drugs that kill rapidly growing human cells.

DEHYDRATION Loss of fluid and salts usually by vomiting and diarrhea.

DISSEMINATE To spread around the body.

DNA Deoxyribonucleic acid. It forms the genetic code.

DYSENTERY Diarrhea due to inflammation of the large bowel.

ECCHYMOSIS Large black or purple areas involving whole limbs due to leakage of blood into the skin.

EDEMA Swelling of tissue or skin due to fluid accumulation.

ENCEPHALITIS Inflammation of the brain.

ENDEMIC A disease regularly present in an area.

ENTERO- Relating to the intestine.

EPIDEMIC Usually, a rapid upsurge in cases of disease in an area.

ERYTHEMATOUS Red.

ESCHAR A black, firmly adherent scab usually associated with anthrax or a tick bite.

EUKARYOTE A cell with a nucleus and internal organelles.

FEBRILE Relating to fever.

FLACCID PARALYSIS Inability to move limbs because the muscle will not contract.

GINGIVOSTOMATITIS Inflammation of the gums and lips.

GISA Glycopeptide insensitive *Staphylococcus aureus*. A new superbug not predictably sensitive to any antimicrobial.

GRANULOMA A chronic red raised or ulcerated lesion that does not heal easily.

HEMATEMASIS Vomiting blood.

HEMATURIA Passing blood in the urine.

HEMOLYSIS Destruction of red blood cells.

HEPATITIS Inflammation of the liver.

IN VITRO In an artificial environment outside the body.

ISOLATE A pure strain of a bacteria or fungus.

JAUNDICE Yellowing of the skin and whites of the eyes.

LARVICIDE A chemical that kills the larval stage of insects, especially mosquitoes.

LEUKEMIA Malignancy (cancer) of white cells in the bone marrow.

LIVE ATTENUATED VACCINE A vaccine comprising the original live pathogen that has been weakened so that it can no longer cause disease, but induces immunity.

LRI Lower respiratory tract infection (below the voicebox).

LYMPHADENOPATHY Swelling of the lymph nodes around the body.

LYMPHOMA Malignant tumor involving lymph nodes.

MACROPHAGE A large dedicated microbe-killing defense.

MACULE A flat spot on the skin.

MALAISE Feeling of being generally unwell and lacking energy.

MECHANICAL TRANSMISSION Transmission of infection by insects where the microbe does not multiply in the insect.

MELENA Passage of black tarry feces due to the presence of blood. The bleeding usually originates in the upper bowel.

MENINGITIS Inflammation of the membranes (meninges) that surround and protect the brain and spinal cord.

MICRON (μ) One millionth of a meter.

MRSA Methicillin resistant *Staphylococcus aureus*. Shorthand for resistance to every penicillin and cephalosporin available.

MUCOSA Mucous membranes.

MUTANT Progeny (offspring) that have a genetic code different to that of the parental microbes.

MYALGIA Muscle aches and pains.

MYCELIUM A meshwork of fibers produced by fungi.

MYOCARDITIS Inflammation of the heart muscle.

NANO- One thousand millionth part of a measurement.

NASO- Relating to the nose.

NECROSIS The localized death of cells caused by, for example, interruption of the blood supply (and oxygen) to those cells.

NEUTROPHIL Professional microbe-killing white cells circulating in the bloodstream.

NORMAL FLORA Microorganisms that normally inhabit organs and other parts of the body.

NUCLEUS A structure within a living cell that contains the cell's hereditary material and controls its growth and reproduction.

OLIGURIA Passage of only small volumes of urine and infrequently.

OOCYST The egg form of protozoa.

OOPHORITIS Inflammation of the ovaries.

ORCHITIS Inflammation of the testicles.

ORGANELLE A specialized, differentiated structure within a cell.

ORO- Relating to the mouth.

PANDEMIC An infection that spreads around the world affecting much of the population.

PAPULE A raised spot on the skin.

PATHOGENESIS The process by which a microbe produces disease.

PERTUSSIS Whooping cough.

PETECHIAE Small pin-point purple or black spots; a result of blood leakage into the skin.

PHOTOPHOBIA A condition where light causes headaches.

PNEUMONIA Inflammation of the lungs' small air sacs (alveoli), where oxygen is normally taken into the body.

PRESENILE DEMENTIA Failure of higher mental function not associated with old age.

PRION An infectious agent made from protein.

PROKARYOTE A cell with no nucleus or internal organelles.

PURPURA Bruise-like spots on the skin.

PUSTULE A pus-filled spot on the skin.

RNA Ribonucleic acid—an intermediate message between DNA and protein production.

SAPROPHYTE A microbe that lives on decaying vegetation or soil.

SEPTICEMIA The body's reaction to viruses or bacteria in the blood.

SINUS An abnormal duct leading from a pus-filled cavity to the surface of the skin.

SPASTIC PARALYSIS Inability to move because of permanently contracted muscles.

SPINAL CORD Large bundles of nerves that transmit impulses to and from the brain.

SPORE A survival package used by some bacteria.

SYNDROME A collection of a number of features associated with a disease.

TACHYCARDIA Fast heartbeat.

TACHYZOITE A rapidly growing form of *Toxoplasma gondii*.

TETANOSPASMIN Toxin released by *Clostridium tetani* that causes prolonged painful muscle contraction.

TOXIN Molecules released by bacteria and fungi to cause damage at sites close to or distant from the microbe.

TRACHEITIS Inflammation of the trachea (windpipe).

TRANSMISSIBLE SPONGIFORM ENCEPHALOPATHY A disease that is transmissible and causes the formation of holes in brain matter and progressive loss of mental function.

UMBILICATED A blister with a dimple in its surface.

UREMIA Urea accumulation in the blood, indicative of kidney failure.

URETER The tubes connecting the kidney to the bladder.

URETHRA The tube that allows passage of urine from the bladder.

URI Upper respiratory tract infection.

VERRUCA A wart on the soles of the feet.

VESICLE A small blister on the skin.

VIRAEMIA Viruses in the bloodstream.

VIRULENCE FACTOR Property of molecules produced by microbes to cause disease.

ZOONOSIS An infection naturally transmitted between vertebrate hosts and humans.

index

picture credits

All images, including jacket,
courtesy of Professor Tony
Hart except: pages 7, 30, 54,
55, 58, 60, 99, 100, 126, 146,
168, 170, 173, 174, courtesy of
Visuals Unlimited/Mediscan;
pages 20, 58, 107, 128, 150,
157, 165, courtesy of SPL;
page 167, courtesy of
www.doctorfungus.org